The Nation's #1 Educational Publisher

McGraw·Hill
Learning Materials

SPECTRUM

GEOGRAPHY

REGIONS OF THE U.S.A.

Grade 4

Authors

James F. Marran
Social Studies Chair Emeritus
New Trier Township High School
Winnetka, Illinois

Cathy L. Salter
Geography Teacher
Educational Consultant
Hartsburg, Missouri

McGraw·Hill
Learning Materials

8787 Orion Place
Columbus, OH 43240-4027

The McGraw·Hill Companies

EAN

9 781577 681540

90000

Program Reviewers

Bonny Berryman
Eighth Grade Social Studies Teacher
Ramstad Middle School
Minot, North Dakota

Grace Foraker
Fourth Grade Teacher
B. B. Owen Elementary School
Lewisville Independent School District
Lewisville, Texas

Wendy M. Fries
Teacher/Visual and Performing Arts Specialist
Kings River Union School District
Tulare County, California

Maureen Maroney
Teacher
Horace Greeley I. S. 10 Queens
District 30
New York City, New York

Geraldeen Rude
Elementary Social Studies Teacher
1993 North Dakota Teacher of the Year
Minot Public Schools
Minot, North Dakota

Photo Credits

2, (tr) Earth Imaging/Tony Stone Images, (br) Aaron Haupt/SRA/McGraw-Hill; **14,** (bc) National Geographic Image Collection, (r) Andy Sacks/Tony Stone Images; **15,** (tc) Martin Rogers/Tony Stone Images, (l) Corbis-Bettmann; **26,** (tr) Randy O'Rourke/The Stock Market, (br) Kunio Owaki/The Stock Market; **27,** (tr) Courtesy of Longfellow's Wayside Inn, Sudbury, Massachusetts; (br) Chris McLaughlin/The Stock Market; **32,** (tr) Clyde H. Smith/Tony Stone Images, (br) Bill Lea; **33,** (tr) Corel Corporation, (br) Joseph Pobereskin/Tony Stone Images; **38,** (tr) Joe Sohm/Chromosohm/The Stock Market, (br) David Barnes/The Stock Market; **39,** (tr) Geri Engberg/The Stock Market, (c) Hugh Sitton/Tony Stone Images, (bl) Jose Fuste Raga/The Stock Market; **50,** (br) Superstock; **51,** (tr) Joe Sohm/Chromosohm/The Stock Market (br) Toyohiro Yamada/FPG International; **56,** Doris DeWitt/Tony Stone Images; **57,** Harald Sund/The Image Bank; **62–63,** UPI/Corbis-Bettmann; **68,** David Hiser/Tony Stone Images; **69,** Kunio Owaki/The Stock Market; **74,** (bc) Dugald Bremner/Tony Stone Images, (r) Kunio Owaki/The Stock Market; **80,** David L. Brown/The Stock Market; **81,** Blaine Harrington III/The Stock Market; **86,** (br) David Meunch/Tony Stone Images, (tr) Ulf Wallin/The Image Bank; **98–99,** Bob Abraham/The Stock Market; **98,** (br) Soames Summerhays/Photo Researchers; **99,** (br) Ben Simmons/The Stock Market; **104–105,** (b) Ken Graham/Tony Stone Images; **105,** Geoffrey Clifford/The Stock Market; **125,** Gary J. Benson/Comstock.

McGraw-Hill
Consumer Products

A Division of The McGraw-Hill Companies

Copyright © 1998 McGraw-Hill Consumer Products.
Published by McGraw-Hill Learning Materials, an imprint of McGraw-Hill Consumer Products.

Printed in the United States of America. All rights reserved. Except as permitted under the United States Copyright Act, no part of this publication may be reproduced or distributed in any form or by any means, or stored in a database or retrieval system, without prior written permission from the publisher.

Send all inquiries to:
McGraw-Hill Learning Materials
8787 Orion Place
Columbus, OH 43240-4027

ISBN 1-57768-154-1

Table of Contents

Introduction to Geography

Lesson 1

Introduction to Regions

Lesson 2

Lesson 3

New England

Lesson 4

Lesson 5

The Middle Atlantic

Lesson 6

Lesson 1

Images of Earth

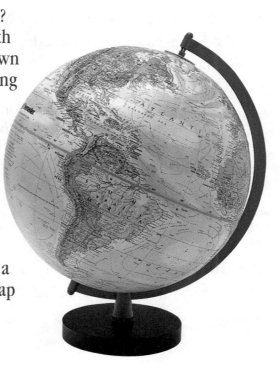

As you read about Earth, think about different ways that we can represent what Earth looks like.

Which one of these three images represents Earth? All of them! The first is an actual photograph of Earth taken from space. The blue areas are water. The brown areas are land. And the white areas are clouds floating above Earth.

The second picture shows a globe. Notice that the shape of the globe matches the shape of Earth shown in the photo taken from space. In this way, a globe represents Earth as it looks from space. A globe gives a more accurate image of Earth because it is shaped like Earth.

The third image shows a map of the world. A map is a flat drawing of some or all of Earth's surface. The map shows how the land is divided into continents. On a flat map, distances or directions can look different than they really are.

When we study geography, we need to be able to "see" Earth's surface. Of these three ways to represent Earth, maps are probably the most useful. A map can be folded up and easily carried around, unlike a globe. And, unlike a photo of Earth from space, a map can show details of Earth's surface. For example, different types of maps show us different things and can be used for different purposes. We might use a map to help us find a particular mountain, river, town, or city. Airplane pilots and ship captains use maps to find their way from place to place. Other types of maps can tell us about population, climate, cultures, or roads.

A map's title tells us what the map shows. Other features of a map may be represented in different colors. Still others may be represented by a symbol, such as a circle, dotted line, or road sign. Sometimes the meaning of the symbols is obvious. For example, a forest may be indicated by a tree. An airport's location might be shown with an illustration of an airplane. To make sure that everyone understands what the symbols mean, a mapmaker includes a legend, or key. This is a box that shows what each symbol or color represents.

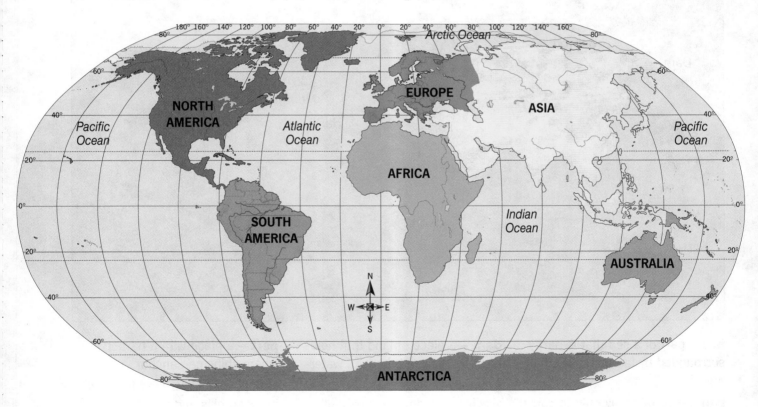

Look again at the images of Earth. Notice that two basic features make up the surface of Earth. They are large masses of land and even larger bodies of water. Do you know what these land masses and bodies of water are called? If you said continents and oceans, you're right!

Lesson 1
MAP SKILLS
Using a Map to Identify Different Forms of Land and Water

Words like mountain, river, or lake describe specific geographic features on Earth's surface. The map below shows and defines several types of landforms and water forms.

bay – part of a sea that is partly surrounded by land, smaller than a gulf

canal – a waterway built to carry water from one place to another

canyon – a deep, narrow valley that has steep sides

coast – land along the sea

dam – a piece of land that holds back the flow of a body of water

glacier – a huge mass of ice that slowly slides down a mountain

harbor – a deep body of water where ships can anchor

mouth – the place where a river empties into a larger body of water

port – a place with a harbor where ships can anchor

valley – the lower land that lies between hills and mountains

volcano – an opening in Earth's crust from which ashes and hot gases flow

waterfall – a stream that flows over the edge of a cliff

4

A. Read the definition of each word and look at its picture on the map. Notice that some words, such as *valley*, relate to land features. Some words, such as *bay*, relate to water.

1. Complete the chart below by writing the terms from the map in the appropriate column. An example has been done for you.

Land	Water
valley	bay

2. Under which of the above columns would you list continents? _____

3. Under which of the above columns would you list oceans? _____

B. Some of the different forms of land and water on Earth's surface are physical, or natural, features. Others are human-made features. Using the map and definitions shown, answer the following questions.

1. Which of the landforms and water forms are physical, or natural, features?

_____

2. Which ones are human-made features?

_____

3. Are there any that can be both formed by nature and made by humans?

_____

Lesson 1
ACTIVITY
Identify land, water, natural, and human-made features.

Name That Feature

The **BIG** Geographic Question

How can we learn to identify specific landforms and water forms?

In the article you read about the different ways Earth's land and water features can be represented. In the map skills lesson you learned the names of some specific land and water forms. Now create and play a game called "Name That Feature."

A. List the seven continents.

1. _____
2. _____
3. _____
4. _____
5. _____
6. _____
7. _____

B. List the four oceans.

1. _____
2. _____
3. _____
4. _____

C. Unscramble each set of letters below to write the names of landforms and water forms.

1. coolvan _____

2. trop _____

3. mad _____

4. aby _____

5. laacn _____

6. coats _____

7. houtm _____

8. tellarfaw _____

9. clearig _____

10. lavley _____

11. cannyo _____

12. borrah _____

D. Using index cards and all of the words you listed in steps A, B, and C, make cards for a "Name That Feature" game.

1. Make cards for each of the land and water features. On one side of the card, write the name and definition of the feature. On the reverse side of the card, draw a picture of the feature.

2. Make cards for the seven continents by drawing their shapes on one side of the card and writing their names on the reverse side.

3. Make cards for the four oceans by writing a description of the continents they are located near on one side of the card and writing their names on the reverse side.

E. Play the game by turning the picture-side of the cards face up and trying to name the feature. Turn the card over to check your accuracy. Challenge yourself by sorting all of the cards into land or water groups.

Lesson 2
States UNITED

As you read about the similarities and differences in various areas of our country, think about how people have grouped similar places to create regions.

The United States is the fourth-largest country in the world in area, and the third largest in population. The features that make up our country range from hot, sandy beaches to cold, snowy mountains, from level prairies to rolling countryside. There are deserts, canyons, rivers, and plains.

The natural resources of the United States are varied, also. Fertile farmland, thick forests, valuable minerals, and abundant sources of water help make our nation one of the richest countries in the world. Our human resources—the United States population—are also diverse. The people who live in the United States have different histories, cultures, and ways of earning a living.

In studying Earth's land and people, it is useful to identify and organize areas of Earth's surface in different ways. To do this, people have created **regions,** areas of land that share certain features that make them different from other areas. A region might be as small as a neighborhood or as large as a country.

Political Regions of the United States

Some maps divide the United States into the Southwest, West, Midwest, New England, and Southeast regions. Other maps divide the United States into the Northeast, Mid-Atlantic, Southeast, Midwest, Southwest, and Mountain/Intermontane West regions. Still other maps show even more and different types of regions.

This map shows one way the United States is divided into regions.

The two basic kinds of regions are *physical* and *human*. **Physical regions** are areas of the natural landscape that might be defined by their landforms, plant and animal life, natural resources, or climate. **Human regions** are based on characteristics of the area's people, such as their language or their government.

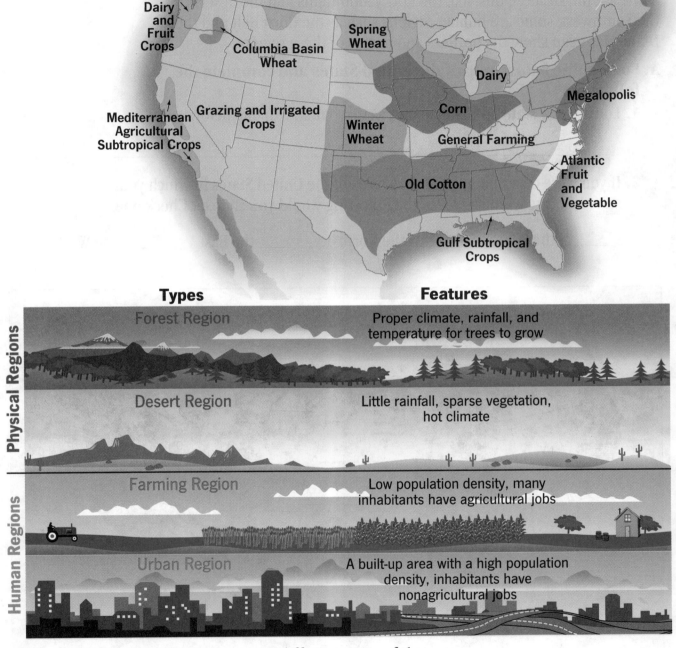

Agricultural Regions of the United States

Pacific Dairy and Fruit Crops

Columbia Basin Wheat

Spring Wheat

Dairy

Megalopolis

Mediterranean Agricultural Subtropical Crops

Grazing and Irrigated Crops

Winter Wheat

Corn

General Farming

Atlantic Fruit and Vegetable

Old Cotton

Gulf Subtropical Crops

Types	Features
Physical Regions	
Forest Region	Proper climate, rainfall, and temperature for trees to grow
Desert Region	Little rainfall, sparse vegetation, hot climate
Human Regions	
Farming Region	Low population density, many inhabitants have agricultural jobs
Urban Region	A built-up area with a high population density, inhabitants have nonagricultural jobs

Defining a region helps us to compare different parts of the United States and to understand what life is like there. It gives us a better picture of the physical and human features of an area. Can you see why creating regions is useful?

Lesson 2
MAP SKILLS
Using an Outline Map
to Create Regions

You know that the United States is divided into several major regions. Some are physical regions, like the Rocky Mountains and the deserts of the Southwest; some are human regions such as the Corn Belt and the Dixie. Maps can show the borders and names of these regions.

A. Look at the below map of the United States and complete the following.

  1. Locate your state on the map and write your state name below.

  _____

  2. If you were describing the general area of the United States in which your state is located, which one of the following would you choose? Check one.

  _____ the South _____ the North _____ the East _____ the West

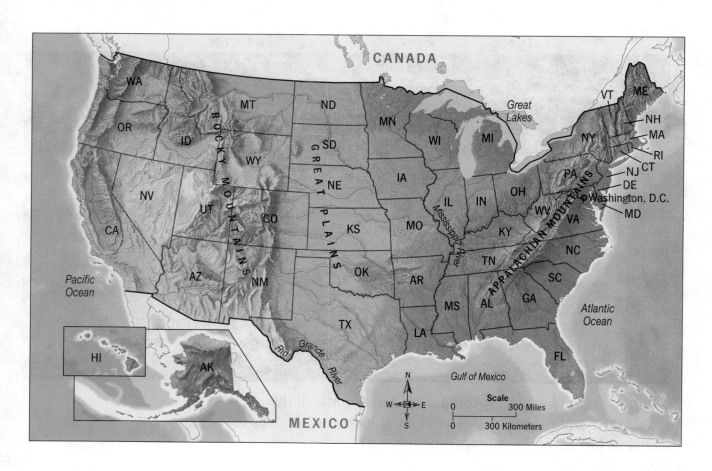

B. Think about where you live in the United States.

1. Write the names of the states near your state that you think have similar features—such as beaches, mountains, forests, or weather—to your state.

_____   _____

_____   _____

2. If you could give your state, together with the ones you listed above, a "region" name, what would you call the region? Think about features the states share and write a region name below.

_____

C. Looking at the map, think about what you know about the different areas of our country. What makes each area special or unique? How would you divide the United States into regions?

1. Write the names of the regions you would include.

a. _____   d. _____

b. _____   e. _____

c. _____   f. _____

2. Make notes about where you will draw the borders for each region.

_____

_____

_____

_____

D. Draw and label your regions on the map on page 10.

E. Compare your map with the regional maps in the article.

1. Does your map have the same number of regions? _____

2. Which, if any, borders are alike, and which are different? Why do you think the same or different borders were chosen?

_____

_____

Lesson 2

ACTIVITY   Identify features of regions.

What Makes Up a Region?

The **BIG** Geographic Question

What are the characteristics of different physical and human regions of the United States?

From the article you learned that there are different ways to divide the United States into regions. The map skills lesson allowed you to create your own regions. Now identify the characteristics of specific human and physical regions of the United States and explain the similarities that make them a region.

A. Use the information in the article and Almanac to help you brainstorm a list of different kinds of regions.

1. Write the names of as many physical regions as you can find. For example, a physical region could be based on temperature.

_____

_____

_____

_____

2. Write the names of as many human regions as you can find. For example, a human region could be based on farming.

_____

_____

_____

_____

_____

B. Choose any one of the physical or human regions from the article, map skills lesson, or this activity to read more about. Then complete the chart below. Write the name of your region in the space at the top. Show your region's location by outlining it on the map. List some features you think are characteristic of your region.

Region

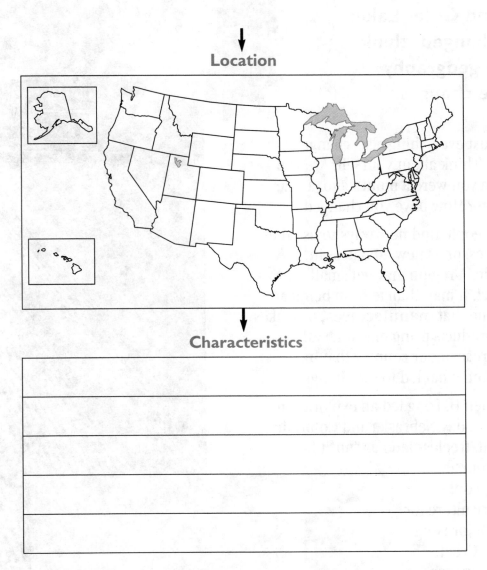

Location

Characteristics

C. Get together with a partner and describe your region aloud. Include characteristics of your region and similarities to other regions. Challenge your partner to identify it and tell whether the region is physical or human. Then compare your maps. How are they similar? How are they different? Do your regions overlap?

Lesson 3

Time Changes Things

As you read about how the Great Plains and Great Lakes regions have changed, think about the role geography played in those changes.

Regions, like almost everything else, change with the passing of time. Think about yourself. Are you different today than you were a month ago, a year ago, or five years ago? How have you changed?

Regions change as people find new ways to use the land and its resources, or as new advances in technology are made. A region may get smaller or larger or break apart. It may change from being a farming region to one that **manufactures,** or makes a large number of products using machinery. Let's take a look at two areas of our country that have changed over time and what led to the changes.

In 1820 Major Stephen H. Long led an exploration party through present-day Nebraska and Colorado. He described the flat, treeless land as "unfit for cultivation, and of course uninhabitable by a people depending upon agriculture for their subsistence." Major Long named the area the "Great American Desert." This area included the present-day states of North Dakota, South Dakota, Nebraska, Kansas, Oklahoma, Montana, Wyoming, New Mexico, Colorado, and Texas within its boundaries.

Was the land uninhabitable? It may have been at first. But, because of advances in agriculture, such as the steel plow, the land indeed became inhabitable. This farming tool made it possible for people to till the land. Over time, further advances in farming tools were made, such as the tractor and combine. Now the part of the Great American Desert known as the Great Plains is one of the world's greatest wheat-producing regions. Another part of the Great American Desert, called the Sunbelt, is now home to many retired people. These retired people settle there because of the warm climate and attractive standard of living.

Other regions that have changed are the Northeast and Great Lakes regions. Manufacturing in North America began in the 1840s in southern New England. The rapid speed and large volume of goods being manufactured during this time led to an Industrial Revolution. During this period of high industry, the leading industrial states stretched from Massachusetts to Illinois.

Today manufacturing is still important to the Great Lakes region. Part of the Great Lakes area is sometimes called the "rust belt" because of the decline of some industries. Southern New England's current strength lies in its people's technical skills and in the region's educational institutions. Tourism is also an important source of income for the New England region. Thousands of visitors come each year to see its many historical sites.

Lesson 3
MAP SKILLS Using Maps to Learn About Changes in Regions

Maps can show recent or past information about places and people. We can compare maps from different time periods to see changes over time.

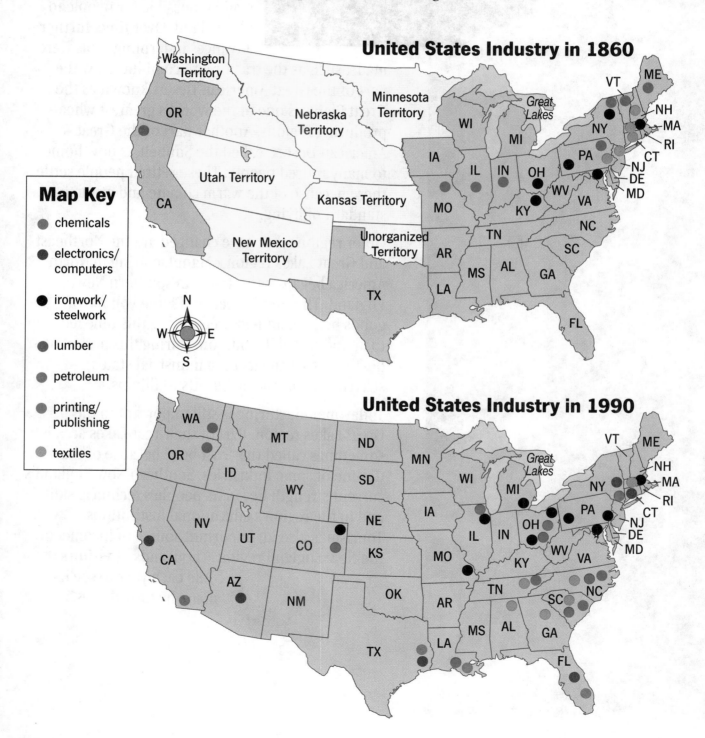

United States Industry in 1860

United States Industry in 1990

Map Key

- chemicals
- electronics/ computers
- ironwork/ steelwork
- lumber
- petroleum
- printing/ publishing
- textiles

A. Look at the map of United States industry in 1860. Answer the following questions.

    1. What were the states involved in industry?

_____

_____

_____

    2. What were the main industries?

_____

_____

B. Compare the 1860 map with the 1990 map. Answer the following questions.

    1. Which states became industry centers in 1990 that were not in 1860?

_____

_____

_____

_____

    2. What types of industries have developed since 1860?

_____

_____

C. What do the maps show you about how regions of industry changed from 1860 to 1990?

_____

_____

_____

_____

Lesson 3

ACTIVITY

Find out the role geography plays in defining regions of the United States.

What A Place Makes

The **BIG** Geographic Question | **In what way does geography affect the types of industries that exist in regions of the United States?**

From the article you learned how two regions of the United States changed over time. The map skills lesson showed you how to use two maps to evaluate changes in the industry of various regions. Now find out how a region's location affects the type of industry that exists there.

A. Look at the maps on page 16 again. List the different types of industry that were in each region in 1990.

1. Northeast and Great Lakes Region

a. _____

b. _____

c. _____

d. _____

2. Mid-Atlantic and Southeast Region

a. _____

b. _____

c. _____

d. _____

e. _____

3. West Coast Region

    a. _____

    b. _____

    c. _____

B. Look at your lists above and answer the following questions.

  1. Near what important physical feature are most of the states with industry
located? _____

  2. Is the physical feature you listed above an important resource needed
to help make many of the goods you listed in Part A?

    Why or why not? _____

    _____

    _____

    _____

  3. Write three of the listed industries that are most important to you.

    _____   _____   _____

**C. Select one of the industries listed above or one in the area where
you live. Research how the industry you selected might be related
to geography. Following are some questions to keep in mind while
doing your research.**

  1. Where is the industry located?

  2. What are the land, water, and resources (including people) in that location?
Are they needed to make the product or good that this industry produces?

  3. Was this industry important in the past? Is it important today? Why?

**D. Make a one-page fact sheet about your selected industry to share
with the class.**

Lesson 4

LAND CHANGING HANDS

As you read about the history of the New England region, think about the many people who have lived in the area and their relationships to the land.

When the first Native Americans arrived in New England long ago, the land was good for growing a number of crops. They grew corn, blueberries, strawberries, beans, squash, grapes, tobacco, and Jerusalem artichokes.

Many years later, Giovanni da Verrazano, Samuel de Champlain, and John Smith were among the first Europeans to step upon and describe the New England coast. Verrazano,

an Italian captain exploring for France, described coastal land that was fertile and cultivated, except for the rocky coast of Maine. Champlain, also exploring for France, mapped and described some of the bays and harbors, especially Plymouth Harbor. English captain John Smith explored Boston Harbor and described the coast and islands as well populated and planted with fruit and corn.

Algonquian
settle in
Eastern
Woodlands
(New England)

Verrazano
explores
east coast

Champlain
explores
Plymouth Harbor

Smith
explores
Boston Harbor

Massachusetts
Bay Colony
settled

| 500s | 1500s | 1524 | 1605 | 1614 | 1620 | 1630 | |

Iroquois form
League of
Five Nations

Pilgrims arrive
at Plymouth
Harbor

When the English settlers arrived on the Mayflower at Plymouth Harbor a few years later, they took over lands inhabited by Native Americans. The English were the first to establish settlements along the coast and inland in the river valleys. Shortly after their arrival, the English fought the French for control of North American lands in the French and Indian War. The English won all the French territory east of the Mississippi River, except for New Orleans.

As the English learned how to successfully farm the land, they grew crops such as corn and tobacco. Later they traded the crops with European buyers. Fishing also provided food for the colonists and income for exporters. The timber industry allowed the English to build wooden ships and houses and to trade lumber. The colonists found that the land provided many resources that were useful for survival.

When the English living in North America grew tired of the control that England had over them, they fought and defeated the British army in the Revolutionary War. A new nation was born. By the mid-1800s many European immigrants were coming to the United States to work in factories. As city populations grew and people had to be fed, farming took on new importance.

The feet of many different people have rested upon the land of New England over time. The land has changed many hands. Each of the different groups of people found ways to use the land that have left a lasting imprint.

Continental Congress adopts Declaration of Independence

First Industrial Revolution

| 1775 | 1776 | 1788 | 1790 | 1791 | 1793 | 1848 |

American Revolution begins

Massachusetts, New Hampshire, and Connecticut become states

Rhode Island enters the Union

Vermont enters the Union

Big numbers of immigrants begin arriving from Europe

Lesson 4

MAP SKILLS
Using a Map to Locate Natural and Human Features

Maps often show physical features, such as landforms and bodies of water. Maps can also show human features—exploration routes and places that people explored, settled, and named.

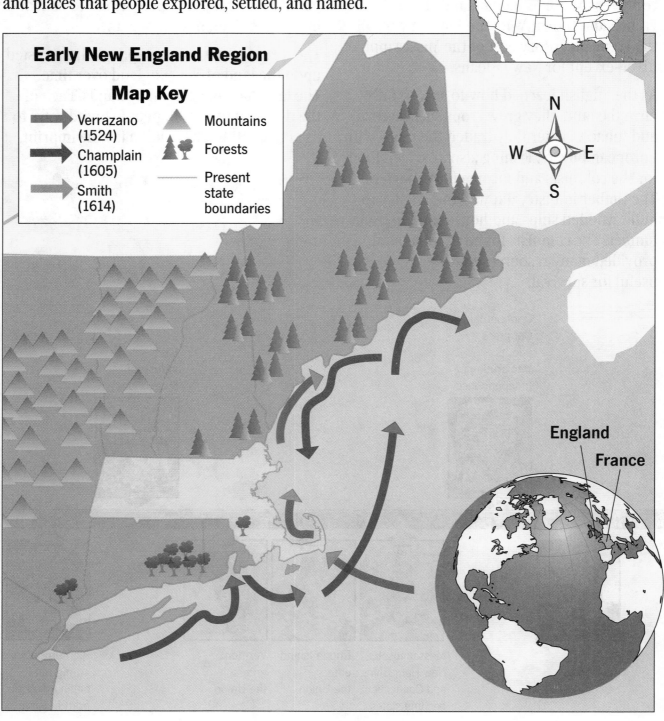

Early New England Region

Map Key

→ Verrazano (1524)

→ Champlain (1605)

→ Smith (1614)

▲ Mountains

▲🌲 Forests

— Present state boundaries

England
France

A. List the six states shown on the map that make up the New England region. Then label them on the map.

1. _____   4. _____

2. _____   5. _____

3. _____   6. _____

B. Use information from the article and Almanac to identify the following. After you identify these places, locate and label them on the map.

1. A name for the region of forests before European exploration

_____

Trace this region on the map.

2. The mountains of eastern North America that run through

New England _____
Draw these mountains on the map.

3. The ocean that the explorers crossed to get to North America _____
Label this ocean on the map.

4. The country that sent Giovanni da Verrazano and Samuel de Champlain

to explore coastal New England _____
Draw a line from this country to the New England region on the globe.

5. The country that sent John Smith to explore coastal New England _____
Draw a line from this country to the New England region on the globe.

6. The harbor that Champlain mapped and described _____
Label this harbor on the map.

7. The harbor that John Smith mapped and described _____
Label this harbor on the map.

Lesson 4

ACTIVITY

Find out how New Englanders have used their land over the years.

Using the Land

The **BIG** Geographic Question | How have different groups in history changed the land in New England?

From the article you learned about the people who settled the New England region. In the map skills lesson you identified the place where European explorers who came to the region landed. Now find out how people used New England's land long ago.

A. Using the chart below, show how each historical group used the land and what it looked like afterwards. Include in your answers what you know about these groups in terms of farming, manufacturing, and the timber industry.

Group of People	Time Period	Ways the Land Was Used	Effects on the Land
Native Americans			
First European Settlers			
Immigrants of the Industrial Revolution			

B. Looking at the information collected on your chart, how has land
use changed over time?

_____

_____

_____

_____

C. What does the **New England** region offer, in terms of its own
industries, that could benefit new businesses? Using the Almanac,
list the industries on the chart below. Then write how you think
each industry would benefit new businesses.

Industry	Benefit

D. Imagine that you are at a meeting of New England governors.
You are to write a letter to businesspeople across the United States.
You are trying to get them to do business in the area. What would
you say to attract them? Write your letter on a separate piece of
paper and share it with the class.

Lesson 5
What's New in New England?

As you read about the New England region now, think about how its past geography compares to its geography today.

Sand dunes at Cape Cod, Massachusetts

Glaciers moving through New England some 12,000 years ago deposited rocks and boulders that made the soil thin and not very suitable for growing crops. These moving sheets of ice deposited ridges of sand that formed Cape Cod.

Native Americans in the New England area long ago found a way to cultivate the rocky land. New Englanders still plant, gather, and fish for foods in ways similar to those first cultivated by Native Americans. Lobster pots, or traps, are dropped in the coastal waters. Lobsters crawl into the pots to eat the bait. More lobsters are caught in Maine than in any other state.

Harbor scene in Maine

26

Buildings, another part of New England's past, still linger. Some of the oldest buildings in the United States are in this region. The Wayside Inn, on the old stagecoach road, in Sudbury, Massachusetts, opened in 1716. It is still open to the public for food and lodging. Elizabeth Portland Head Lighthouse, built in 1790, still warns ships away from the dangerous, rocky shores of Cape Elizabeth, Maine.

Wayside Inn today in Sudbury, Massachusetts

Meetinghouses often served as both a town hall and a place of worship in colonial days. All town meetings and elections were held there. Even today local government is often organized and managed through town meetings, at which every citizen can vote on any local law. In sharp contrast to the meetinghouses of the past are the skyscrapers of today. The 60-story John Hancock Tower can be found in Boston, Massachusetts, the largest city in New England.

New England is known for its autumn foliage. Tourists from many other states drive through the mountains to look at the beautiful red, orange, and gold leaves that have become a vivid image representing New England.

Elizabeth Portland Head Lighthouse in Cape Elizabeth, Maine

Lesson 5

MAP SKILLS Using Letters to Recall Map Elements

Acronyms are words formed by the first letters of a series of words. They are often used to help us remember a list of important things. The elements of a map are easy to recall with the help of an acronym.

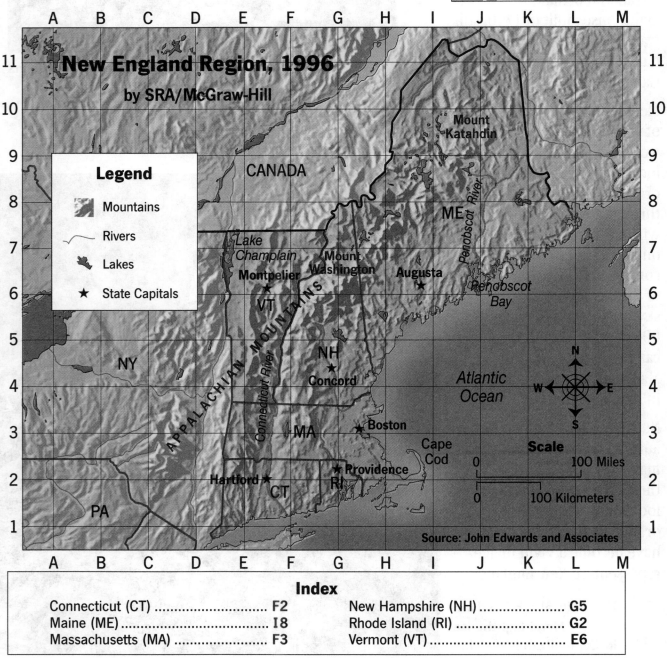

New England Region, 1996

by SRA/McGraw-Hill

Legend

- Mountains
- Rivers
- Lakes
- ★ State Capitals

CANADA

Mount Katahdin

Lake Champlain

Mount Washington

Montpelier

VT

Penobscot River

ME

Augusta

Penobscot Bay

NY

APPALACHIAN MOUNTAINS

NH

Connecticut River

Concord

Atlantic Ocean

MA

Boston

Cape Cod

Hartford ★

CT

★ Providence

RI

PA

Scale

0   100 Miles

0   100 Kilometers

N W E S

Source: John Edwards and Associates

Index

Connecticut (CT) F2	New Hampshire (NH) G5
Maine (ME) I8	Rhode Island (RI) G2
Massachusetts (MA) F3	Vermont (VT) E6

A. Match each of the following words with one of the clues given below and reveal an acronym for map elements. The circled letters form the acronym.

Legend   Grid   Author   Date   Source
Title   Index   Orientation   Scale

1. Tells what the map is showing ◯ __ __ __ __

2. Tells what direction things appear in relation to each other

◯ __ __ __ __ __ __ __ __ __ __

3. Tells when the data was gathered ◯ __ __ __

4. Tells who gathered the data ◯ __ __ __ __ __

5. A box that shows the meanings of the symbols ◯ __ __ __ __ __

6. Used to compare map distance to actual distance ◯ __ __ __ __

7. Lists alphabetically the places shown on the map ◯ __ __ __ __

8. Helps find the location of a particular place on a map ◯ __ __ __

9. Shows who provided the data ◯ __ __ __ __ __

10. What is the acronym? __ __ __ __ __ __ __ __ __

B. Look at the New England regional map. Identify or describe the map elements using the acronym TODALSIGS.

1. T _____

2. O _____

3. D _____

4. A _____

5. L _____

6. S _____

7. I _____

8. G _____

9. S _____

Lesson 5

ACTIVITY
Make a map of the physical and human elements of the New England region.

A Map of New England

The **BIG** Geographic Question

What are some of the key physical and human elements of New England?

From the article you learned about physical and human features of New England. In the map skills lesson you learned an acronym to help you recall map elements. Now make a map showing physical and human elements of the New England region.

A. Look at the features listed on the chart below. Choose at least one of these New England resources to research in the Almanac. Take notes as you do your research.

Feature	What to Research	Almanac Notes
Mountains	name, where located	
Rivers	name, where located, mouth	
Vegetation	forest lands	
Natural Resources	types of natural resources	
Wildlife	types of wildlife	

30

B. Decide how you will show the information you have noted on a map. Use a sheet of scrap paper to plan and sketch your map. Then draw your map of the New England region in the space below. Give your map a title.

Lesson 6

FROM THE MOUNTAINS TO THE OCEAN

As you read about the coastal plains and mountain areas of the five Middle Atlantic states, think about how the physical features of an area affect what people do.

Visit a Pennsylvania Farm!

The Middle Atlantic region is one of great **diversity,** or many differences. Its five states—New York, New Jersey, Pennsylvania, Delaware, and Maryland—offer many choices of scenery and activities.

All of the Middle Atlantic states, except Pennsylvania, border the Atlantic Ocean. This gives them a seashore, which is great for swimming and even better for commercial fishing. The flat stretch of land along the shore is a **coastal plain.** Although Pennsylvania does not border the Atlantic Ocean, it does have a coastal plain. In fact, Philadelphia, Pennsylvania, is a major shipping port because it's just a short float down the Delaware River through the Delaware Bay off the Atlantic Ocean.

Hike the Appalachian Mountains

Water dominates the eastern shores of the Middle Atlantic, but go west and you'll find mountains—lots of them! The Appalachian Mountains await you. These scenic mountains have dramatic elevations that made traveling and trading difficult for early settlers in the United States.

In 1817 American engineers began building the 363-mile Erie Canal to connect the Great Lakes with the Hudson River. Eight years later, the canal was completed and was the longest uninterrupted canal in the world. The Erie Canal provided quicker and easier travel and trade to the new states beyond the Appalachian Mountains.

Today highways and railroads connect the mountains with the seashore and people with the eastern and western cities. So whether you would like to hike the Blue Ridge Mountains or sail on the Atlantic Ocean, the Middle Atlantic region has just the place for you.

Lesson 6
MAP SKILLS Using an Elevation Key

We know that the land that a map shows us is not always flat. We use a map's **elevation key** to help us understand the shape of the land. **Elevation,** or **relief,** is measured in meters or feet. It is the height the land rises above sea level.

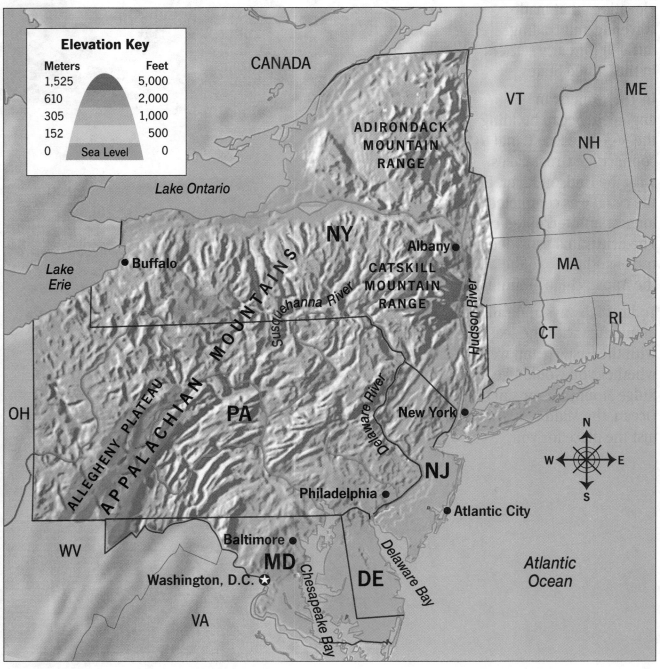

Elevation Key

Meters		Feet
1,525		5,000
610		2,000
305		1,000
152		500
0	Sea Level	0

CANADA

VT

ME

NH

ADIRONDACK
MOUNTAIN
RANGE

Lake Ontario

NY

MA

Albany

CATSKILL
MOUNTAIN
RANGE

Lake
Erie

Buffalo

Hudson River

RI

CT

MOUNTAINS

Susquehanna River

OH

APPALACHIAN

ALLEGHENY PLATEAU

PA

Delaware River

New York

NJ

N

W   E

S

Philadelphia

Atlantic City

WV

Baltimore

MD

DE

Delaware Bay

Washington, D.C.

Chesapeake Bay

Atlantic
Ocean

VA

A. Look at the elevation key to answer the following questions.

1. Write the elevation of the land shown in yellow.

_____ meters   _____ feet

2. Which elevation is higher, the light green area or the orange area?

_____

B. Look at the map of the Middle Atlantic region.

1. What is the elevation of most of the land along the Atlantic Ocean?

_____

2. Describe how the land changes as you travel from west to east.

_____

_____

C. Find Buffalo, New York, on Lake Erie. Draw a line from west to east going from Buffalo to Albany, New York.

1. What two colors do you pass through? What are the elevations?

a. Elevation color   **b.** Elevation in feet

_____   _____

_____   _____

2. What is the name of the mountain range north of the line that you drew?

_____

3. Continue your line from Albany, New York, by tracing the Hudson River south to New York City. What is the elevation in feet? _____

4. Do you think the route from Buffalo to Albany was a good route for the Erie Canal? Why? _____

_____

_____

_____

Lesson 6

Activity Find out how the geography of a region influences where people live.

Why Live There?

The **BIG** Geographic Question | **How do the land and water of a region affect settlement patterns?**

In the article you read about the diverse physical features of the Middle Atlantic region. In the map skills lesson you looked at the elevation of this region. Now find out more about how geography affects people's locations and occupations.

A. Look in the Almanac for the population density map for the five states of the Middle Atlantic region.

1. Where is the population the greatest? Why?

_____

2. Which two states have the greatest rural areas?

_____   _____

3. What ocean is to the east of this region? _____

4. What two Great Lakes border the western side of this region?

_____   _____

5. List three major rivers in this region.

_____   _____   _____

6. List two big bays on the east coast.

_____   _____

7. List mountains in this region.

_____

B. Think about the geography and the population of the Middle Atlantic region.

1. What do you notice about the relationship between the features of the land and where people live? _____

_____

_____

_____

2. How do you think people decide where to live? _____

_____

_____

_____

_____

C. Now, using clay or papier-mâché, make a model of what you have learned about the Middle Atlantic region. Be sure to include the major landforms and waterways and mark the cities.

Here are a few hints:

- Glue yarn or string to your model to mark the major waterways.

- Use extra clay or papier-mâché to build up the mountain areas.

- Use flags made of toothpicks and small pieces of paper to mark the mountain ranges. Write the elevations on your flags.

- Make your own map key and symbols to show information.
  See an example below.

Sample Map Key

★ state capital

◯ ocean

▲ mountains

Lesson 7

BosWash MEGA What?

As you read about the BosWash Megalopolis, think about how its geography makes it a different kind of region.

BosWash Megalopolis is the name given to the urban region that stretches along the Atlantic Ocean from north of Boston to south of Washington, D.C. It contains five major urban regions along the Middle Atlantic and southern New England coast—Boston, New York City, Philadelphia, Baltimore, and Washington, D.C.

Megalopolis comes from Greek words—*megas* means "great" and *polis* means "city." A megalopolis contains a number of cities and other urban areas where boundaries overlap. Its population density is so great that it appears to be one big city. At the center of the BosWash Megalopolis is New York City.

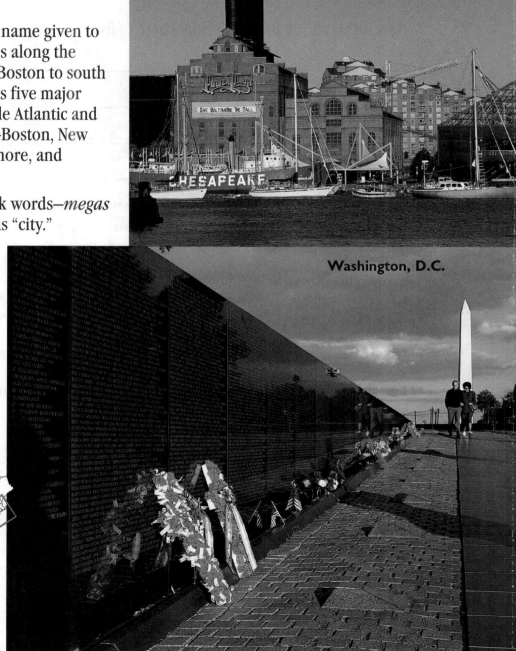

Baltimore

Washington, D.C.

The **BosWash Megalopolis** is a region of striking contrasts.

 It has rural areas that provide the cities with farm products and recreation.

 It has people from many different cultures.

 It is an area of great wealth and great poverty.

 It has fine museums and symphonies, but also areas with high crime rates.

 It has harbors, islands, sky-scrapers, and factories.

Boston

New York City

Philadelphia

The location of the BosWash Megalopolis is on the Atlantic Ocean, a busy ocean route between the United States and Western Europe. Most cars, ships, trains, and airplanes arrive and leave the United States at this "great city." Within this megalopolis you have some of the world's greatest harbors and ports, shores and coasts, population densities and diversity, and industry and trade all rolled into one great area.

Lesson 7
MAP SKILLS Using Maps to Understand the BosWash Megalopolis

We can use a series of maps to see where a place is located within a larger area. The first map below shows New York City, one of the urban areas located in the BosWash Megalopolis. The second map shows where the BosWash Megalopolis is located in the larger New England and Mid-Atlantic regions. The third one shows where these regions are located in the United States.

Map 2

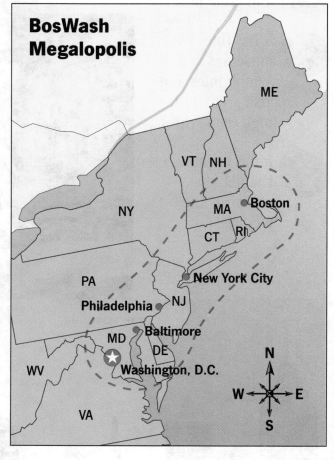

BosWash Megalopolis

Map 1

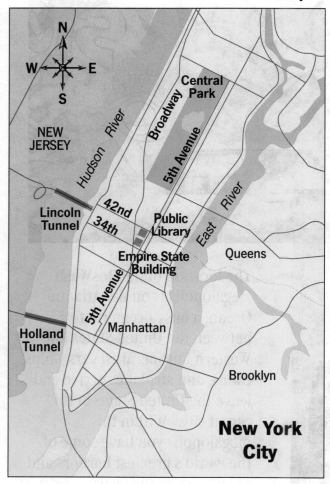

New York City

Map 3

1. What is the name of the closest major city southwest of New York City? _____

   _____

   Which map showed you New York City with other major cities? _____

2. What building is located slightly southwest of the New York Public Library? _____

   _____

   Which map shows the building? _____

3. In what region is the state of New York located? _____

   Which maps show you the state of New York in a region? _____

4. What street runs along the west side of Central Park? _____

   Which map shows the street? _____

5. Look at the three maps and answer the following questions.

   a. What river borders the east side of Manhattan? _____

   b. What river borders the west side of Manhattan? _____

   c. Which one of these rivers has two tunnels connecting New York and

      New Jersey? _____

   d. Name three boroughs, or sections, within New York City that are
      shown on the map.

      _____

      _____

      _____

6. Put an *X* in the box next to the map that shows a larger part of the United States.

   [ ] Map 1   [ ] Map 2   [ ] Map 3

Lesson 7

ACTIVITY
Compare and contrast the five major
urban regions of the BosWash Megalopolis.

Megalopolis Mania

The **BIG** Geographic Question | **How are the urban regions of the BosWash Megalopolis alike and different?**

From the article you learned what the BosWash Megalopolis is. The map skills lesson showed you New York City and the megalopolis on different scale maps. Now compare and contrast the five major urban areas that make up the BosWash Megalopolis.

A. On the chart below, write down some things you know about each urban area of the BosWash Megalopolis. Use the article and map skills lesson to help you with this information. One urban area has been done for you.

Megalopolis Urban Area	Where It's Located	Physical Features
Washington, D.C.	between Virginia and Maryland	• near Potomac River

B. Using the Almanac information, complete the following to learn how the megalopolis urban areas are alike and connected.

1. All five urban areas have populations

_____ under 500,000   _____ over 500,000.

2. All five urban areas have _____ airports that connect

them with each other and cities around the _____.

3. All five urban areas are located near waterways, such as the

_____ Ocean or a _____.

C. Use the information from your chart and the Almanac to create a 2- or 3-dimensional cityscape of the BosWash Megalopolis. A cityscape is a representation of a city's landscape. Plan your cityscape on a scrap piece of paper. Include all five urban areas and show that they are connected to form one region. Here are some suggestions.

- Draw and color or paint different kinds of landscapes on posterboard to represent each urban area's scenery. The scenery might include highways, railways, airports, seaports, and so on.

- Use index cards to draw and cut out the faces of features such as buildings, trees, or airplanes.

- Prop up all of your features by taping one end of a piece of cardboard against their back sides and the other end to the landscape.

- Write a description of the BosWash Megalopolis. Include information about each urban area's physical features and the transportation routes that connect them.

_____

_____

_____

_____

_____

_____

_____

Lesson 8

Shrinking Swamps!

As you read about the fragile and unique swamps of southern Florida, think about the effects humans have had on these wetlands environments.

The Everglades in southern Florida is an area known as a "River of Grass." It has this nickname because the ground is always covered with a sheet of shallow water and grasses, shrubs, and trees. The shallow layer of water makes the area a **wetlands** region. Wetlands are areas, such as swamps and marshes, that have wet soils. The Big Cypress Swamp is another wetlands environment in southern Florida. It, too, is a low-lying wetland area with plant life poking through. **Watershed** is a feature common to these areas. It is all the water that eventually drains from a land area into a creek, lake, or river.

The Everglades and the Big Cypress Swamp are very fragile areas. This means they can easily be destroyed. The Everglades **ecosystem**—the plants and animals such as alligators, manatees, and the Florida panther that live in the area—is highly dependent on the water that flows through southern Florida. The Big Cypress Swamp shelters snakes, fish, and birds and plants such as cypress trees.

People did not always protect these fragile wetlands environments. For a long time, some people saw these wetland areas as wasteland ready to be used or developed. Other people saw them as places to drive off-road vehicles, which killed vegetation growing there. Some people chopped down the trees in these areas and did not plant new ones. Still other people collected plants and hunted the animals, disrupting the food chains.

When people started to see the abundance of life sustained by the Everglades, instead of preserving it they began draining the water to prepare it for farming. The land was cleared to plant sugarcane. People also built canals that connected Lake Okeechobee to the ocean. Eventually, the land was cleared enough to build cities and roads.

Each change damaged the area. There was a decline in wildlife. Plants and animals, such as numerous varieties of the colorful Florida tree snails, became extinct. Canals and cities changed the water flow and pushed back the boundaries of the swamps. People also brought pollution to the swamps.

Eventually, people began to realize the importance of the wetlands and the effects of their actions. Efforts are now being made to save the wetlands of Florida. The Big Cypress National Preserve and Everglades National Park were created to protect these areas. In addition, groups such as the Nature Conservancy are working hard to keep the lush marsh areas from shrinking even more!

The coastal wetlands in Southern Florida are among the most fragile environments in the United States. They are among the most important environments in Florida because they help control flooding, provide clean water, shelter wildlife, and support food chains. The changes that have occurred in these areas are examples of how human actions can modify physical environments. As people learn more about the abundance of life in these coastal wetlands, they can modify their actions to save these environments.

Lesson 8

MAP SKILLS
Using Maps to Compare Natural and Human-Made Features

Maps are created to show specific information such as roads, state borders, or physical features of the land. They can help us identify which physical features of the land are natural or human-made. These two kinds of physical features can be shown and compared on two maps.

Map A

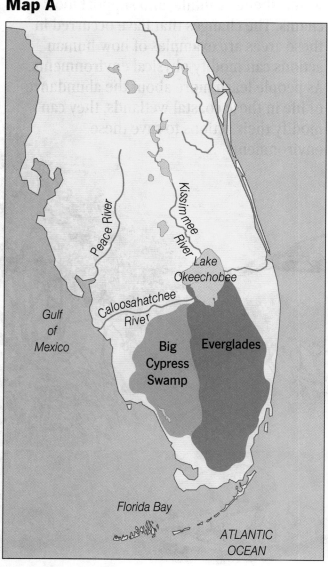

Peace River

Kissimmee River

Lake Okeechobee

Caloosahatchee River

Gulf of Mexico

Big Cypress Swamp

Everglades

Florida Bay

ATLANTIC OCEAN

Map B

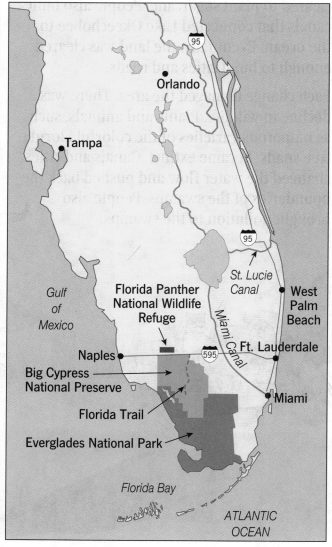

95

Orlando

Tampa

Gulf of Mexico

Florida Panther National Wildlife Refuge

Naples

Big Cypress National Preserve

Florida Trail

Everglades National Park

St. Lucie Canal

95

West Palm Beach

Miami Canal

595

Ft. Lauderdale

Miami

Florida Bay

ATLANTIC OCEAN

A. Look at the maps of southern Florida. Put an *X* on the following places on each map.

Map A
1. Lake Okeechobee
2. Big Cypress Swamp
3. Everglades
4. Kissimmee River

Map B
5. Florida Panther National Wildlife Refuge
6. Everglades National Park
7. Big Cypress National Preserve
8. Florida Trail

B. Use both maps to complete the following.

1. Indicate on which map you can find each of the following and whether each is natural or human-made.

   a. Big Cypress Swamp? _____

   b. Orlando? _____

   c. Big Cypress National Preserve? _____

   d. an interstate highway? _____

   e. a canal? _____

   f. rivers? _____

2. Look at Map B. Where are most of the cities located?

   _____

   _____

3. Draw the information from Map B onto Map A.
   a. Place the cities, roads, and canals on Map A.
   b. Add the boundaries to show Big Cypress National Preserve and Everglades National Park.
   c. Are any human-made features such as cities located inside Big Cypress

   National Preserve and Everglades National Park? _____

4. Now look at Map A. Describe how much of the Everglades and Big Cypress Swamp has been preserved by humans and what has happened to the remainder of the wetlands.

   _____

   _____

Lesson 8

ACTIVITY
Find out the pros and cons of changing fragile wetlands.

Changing Environments

The **BIG** Geographic Question

Should people develop or preserve fragile ecosystems like the wetlands in southern Florida?

From the article you learned that the wetlands of southern Florida are very fragile but important environments. In the map skills lesson you looked at some human-made versus natural features in southern Florida. Now find out what different groups want to do about the future of the wetlands.

A. Indicate what you know about the following wetland features. Add other features from the article or your own research. Write *N* for natural, *H* for human-made, *+* for positive impact on the wetlands, and *–* for negative impact on the wetlands.

Features	Type	Impact
parking lots		
paved roads		
cypress trees		
sawgrass		
canals		
alligator holes		

B. Imagine you have to present the viewpoint of a group that has a special interest in the wetlands of southern Florida. Choose one of the groups below. You might want to talk with an adult friend or family member to see what they know about the groups listed. Best of all, if you know someone who fits one of these descriptions, you might want to ask that person the questions in section **C** below. Circle the group whose ideas you want to represent.

1. Land developers—want to build new condominium complexes or vacation areas
2. Environmentalists—want to preserve the environment
3. City dwellers moving out of the city—want to move to a quiet place to enjoy nature and get away from the crowded city
4. Farmers—want land to farm and water to irrigate crops
5. Senior citizens retiring to southern Florida—want to move to a place with warm weather and plenty of recreational activities close to home

C. See how many of the following questions you can answer about your group. Do research if you need more information.

1. Does your interest group want to preserve or develop the wetlands?

_____

2. What has happened to the wetlands in the past that you think is good?

_____

3. What has happened to the wetlands in the past that you think is bad?

_____

4. What are your group's plans for the wetlands?

_____

5. Why would following your group's plans be a good idea?

_____

D. Present to a friend or family member your group's ideas on the issue of preserving versus changing the wetlands. Use books, pictures, and maps to enhance your presentation.

Lesson 9
From Cotton Belt to Sun Belt

As you read about the South, notice how geographical changes are bringing about a new look for the "Old South."

In the 1800s the South was known as the "Cotton Belt" because cotton was the main crop. Plantation owners put their money into growing cotton, and their profit came from selling it. Whether rich or poor, a free person or a slave, the lives of people in the South depended on cotton. In the "Old South," cotton was king.

Growing cotton did not always go smoothly. Some years plantation owners ruined their soil by growing only cotton. Then they had to buy more land. Other years plantation owners grew too much cotton, and the price of cotton was down. During these years, profits were low. Still other years the Mexican boll weevil destroyed the cotton. Then the plantation owners made no money.

In the 1860s the Civil War brought change to the "Old South." Before the Civil War, plantation owners had slaves to plant and harvest the cotton. After the Civil War, plantation owners could no longer own slaves. They needed lots of workers, but they had no money to pay the workers. Many large plantations were divided into small farms. Freed African American slaves became tenant farmers, paying rent to farm on land that had once been part of their former owner's large plantation.

A field of cotton in Mississippi

By the 1900s people were moving from farms to the cities. People in the cities needed jobs. In the 1930s the Tennessee Valley Authority (TVA) built dams along the rivers that helped provide electricity needed by factories in the region. Between 1900 and 1940 14 million people left farms in the South, including about two million African Americans. The attraction of jobs in factories encouraged many people to move to the Midwest, Northeast, and California. Poor conditions in the South also gave many a reason to leave.

This Tennessee dam produces electricity used by businesses.

Eventually, better ways of communicating and better roads brought businesses to the South. State and local governments offered tax breaks to industries that moved into the area. Between 1940 and 1980 industry in the South doubled, and cities grew. The people who lived in the South could now find jobs in occupations such as food processing, textiles, and the military. The "Old South" was changing, and a "New South" was taking shape.

The South's warm climate began to attract more and more people. The invention of air-conditioning made it possible to work comfortably in large buildings year-round. Some people moved south to retire. Others went to enjoy the warm weather and visit the many beaches and historical sites.

Tourism eventually became one of the big industries of the South. In fact, Atlanta, Georgia, was home to the 1996 Summer Olympics. This event presented the image that the South wants the world to see—a "New South" that is modern, broad-minded, and linked to the world. Today the cotton belt of the "Old South" prefers to be known as the sunbelt of the "New South."

Stone Mountain, in Georgia, is one of the South's many historic sites.

Lesson 9
MAP SKILLS

Using Maps to Study Industry Location Change Over Time

Map A shows where centers of industry were located before the Civil War in the 1860s. Map B shows where centers of industry were located in 1990. We can use these maps to figure out how industrial locations have changed over time.

Major Industrial Centers, 1860

Washington Territory
Oregon
Minnesota Territory
Boston
San Francisco
Nebraska Territory
Chicago
Cleveland
New York City
Philadelphia
Utah Territory
Cincinnati
Pittsburgh
Baltimore
St. Louis
Richmond
Kansas Territory
Louisville
New Mexico Territory
Unorganized Territory

Map Key
● industrial center

Major Industrial Centers, 1990

Seattle
Portland
N W E S
Boston
Chicago
Cleveland
New York City
Philadelphia
San Francisco
Cincinnati
Pittsburgh
Baltimore
St. Louis
Richmond
Louisville
Greensboro
Los Angeles
Nashville
Charlotte
Phoenix
Atlanta
San Diego
Birmingham
Dallas
Jacksonville
Houston
New Orleans
Orlando
Tampa
St. Petersburg
Miami

A. Look at the maps and answer the following questions.

1. In 1860, were most of the cities located in the northeastern states or the

   southeastern states? _____

2. In 1990, where were most of the cities located? _____

   _____

3. What cities were important industrial centers in 1990 that were not important

   in 1860? _____

   _____

   _____

4. To what areas did industry spread in 1990? _____

   _____

B. Looking at the maps and your answers to the above questions, draw some conclusions about how the location of major industrial centers has changed over time.

1. Read the statement: Manufacturing moved from the industrial
   northeast to the "New South." Explain why you agree or disagree
   with this statement.

   _____

   _____

   _____

   _____

2. Why do you think most of the industrial centers in 1860 and 1990 were
   mainly on the east and west coasts?

   _____

   _____

   _____

   _____

Lesson 9

ACTIVITY Find out what makes a good tourist attraction.

"Hot" Tourist Spots

The BIG Geographic Question

Why has tourism become such big business in the Southeast?

From the article you learned about the South and how it has changed. The map skills lesson showed you how the location of industrial centers changed over time. Now find out why tourism has become such a big business in the Southeast.

A. Circle one of the following tourist attractions in the Southeast, or write the name of another southeast destination that you would like to know more about.

- Civil Rights Monument
- Kennedy Space Center
- The Appalachian Trail

- Walt Disney World
- Smoky Mountain National Park
- The Everglades National Park

B. Answer as many of the following questions as you can about the tourist attraction you selected. Use a pencil so you can change your answers later by checking them against the Almanac information.

1. Where is your tourist destination located? _____

2. What is the climate like there? _____

3. What are some specific things to do there? _____

_____

4. What else is nearby that is of interest? _____

_____

5. How could you travel there? _____

54

C. What do you think are important features of a tourist attraction or destination? Use the chart below to indicate your rating by marking a **+** if the feature is important. Mark a **–** if the feature is not important. Then ask two classmates or family members for their ratings.

Feature of Tourist Attraction	Your Rating	Person 1 Rating	Person 2 Rating
places to camp nearby			
places to eat nearby			
cost to get in			
location			
weather and climate			
other tourist sites nearby			
educational			
fun			
variety of activities			
shopping			

D. Design a travel brochure for that tourist attraction. Include pictures, a map, or any other graphic that will add interest to your brochure. The brochure should include reasons why people should visit the site. Use ideas from the chart above. Then look over your travel brochure and those of your classmates. What common feature is important in attracting people to the Southeast?

_____

_____

_____

_____

Lesson 10

HEARTLAND, USA!

As you read about the Midwest region, think about how its geography makes it the "Heartland."

The Midwest includes 12 states that are located in the center of the United States. This is one reason this region is often called the "Heartland" of the country. Another reason is that some people think that Midwesterners still live like early settlers of the United States did—farming the land, living in small towns, and leading a slower paced lifestyle. Many Midwesterners share this view of their region, even though the Midwest has large cities like Chicago, Illinois, and St. Louis, Missouri.

Part of the Midwest region was once covered with wild prairie grass. Now it is covered with grains. The area has rich soil, a good climate for farming, and in most areas, a plentiful supply of water. This makes it an excellent place for growing crops such as corn or raising livestock such as cows. The Corn Belt, the Wheat Belt, and the Dairy Belt are all in the Midwest.

The Midwest is also known as the "industrial heartland" of America. Many major midwestern cities became famous in the 1950s and 1960s for their manufactured products. Detroit, Michigan, has traditionally been known for its automobile industry. Gary, Indiana, has been known for its steel mills. Chicago, Illinois, has been known for its meat-packing industry. Minneapolis, Minnesota, has been known for its flour mills. These products have been transported all over the world by way of the country roads, interstate highways, railroads, rivers, and lakes of the Midwest.

The Midwest has huge cities such as Chicago.

What do you see when you travel around the Midwest today? You see big cities and small towns. You see grain and dairy farms, as well as factories. There are farmers' markets and grocery superstores. You travel interstate highways and gravel roads. You see small airfields and major airports. You meet people from a variety of ethnic backgrounds. You see what many people think of as "typical" Americans—farmers, factory workers, business people, educators, and truck drivers. You see all the reasons the Midwest is called the "heartland" of the United States.

The Midwest has miles of farmland, like these broad, flat plains in Kansas.

Lesson 10
MAP SKILLS
Using a Dot Map to Identify
Grain Belts of the Midwest

Maps can show where crops are grown. Maps can also show how much of an area is planted with different crops. Knowing where a crop is grown and how much of it is grown in an area helps you identify the "belt," or zone, for that crop. Some areas of the Midwest are known as the Corn Belt, the Winter Wheat Belt, and the Spring Wheat Belt.

Corn and Wheat Belts

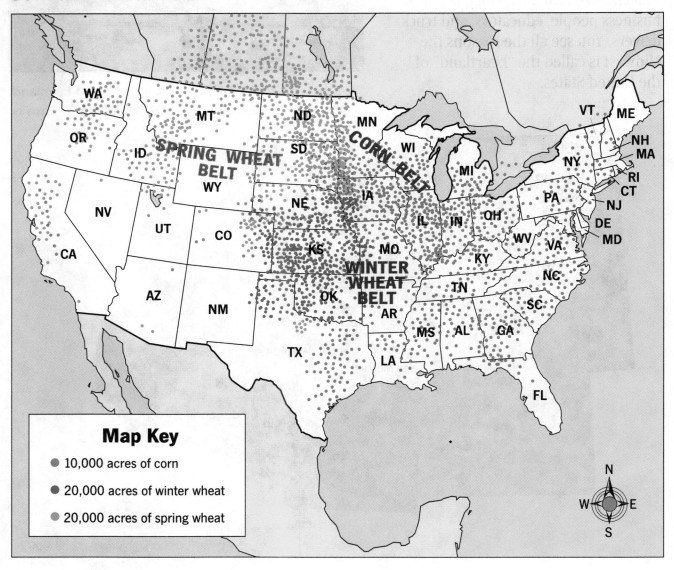

WA
MT
OR
ID
SPRING WHEAT BELT
WY
NV
UT
CO
CA
AZ
NM
ND
SD
NE
KS
OK
TX
MN
CORN BELT
WI
IA
IL
MO
WINTER WHEAT BELT
AR
LA
MI
IN
OH
KY
TN
MS
AL
GA
FL
PA
WV
VA
NC
SC
NY
VT
ME
NH
MA
RI
CT
NJ
DE
MD

Map Key
- 10,000 acres of corn
- 20,000 acres of winter wheat
- 20,000 acres of spring wheat

N
W E
S

58

A. Look at the map and circle the following.

1. Illinois   3. Nebraska   5. Missouri
2. Iowa   4. Arizona   6. Map key

B. Use the map and map key to answer the following questions.

1. a. What does a red dot mean? _____

    b. What does a green dot mean? _____

2. a. Name at least five states where corn is grown. _____

     _____

     _____

    b. How do you know corn is grown in the states you named? _____

     _____

C. Answer the following.

1. How many dots do you see in Arizona? _____

2. What color is each dot? _____

    _____

3. How many acres of corn are grown in Arizona? _____

    How do you know? _____

4. Is more corn grown in Arizona or Illinois? _____

    Explain how you know. _____

    _____

D. Write a few sentences comparing the amount of corn grown in Ohio to the amount of corn grown in Missouri.

_____

_____

_____

_____

Lesson 10

ACTIVITY Find out how a natural disaster affects people—directly and indirectly.

Floods, Farmers, and Food

The **BIG** Geographic Question
How does a natural disaster in one region affect people in another region?

From the article you learned that the **Midwest is known as the industrial and farming heartland of the United States. A dot map in the map skills lesson helped you locate corn and wheat belts in the Midwest. Now find out how something that happens in one region of the United States can touch the lives of people in another region and sometimes all over the world.**

A. List some natural disasters that you think could hurt a farmer's crops.

1. _____   4. _____

2. _____   5. _____

3. _____   6. _____

B. The Great Midwestern Flood of 1993 was a natural disaster. Try to answer as many of the following questions as you can about the flood. Use a pencil so you can change your answers later. Look in the Almanac to check and adjust your answers.

1. List the nine states that were directly affected by the flood. _____

_____

2. Which two rivers were the main source of the flood waters? _____

_____

3. What were some of the crops grown in the flood area? _____

_____

60

C. Many people felt the effects of the Great Flood of 1993. Read each effect. Mark *D* if it *directly* affected the people; mark *I* if it *indirectly* affected the people.

Effect	People in Flood Area	People Outside Flood Area
1. Flood waters and river bottom sand covered farm fields.		
2. Homes and other structures were filled with water and mud.		
3. Roads closed and bridges washed away.		
4. Drinking water was contaminated.		
5. Farmers couldn't harvest crops or plant new crops.		
6. Volunteers helped flood victims.		
7. Certain foods were in short supply in the grocery stores.		

D. Use the information from the article, the map skills lesson, the Almanac, and the information you collected above to create a news broadcast about the flood. Describe the areas directly affected by the Great Midwestern Flood of 1993. Describe the physical features involved and the problems caused by the flood. Write your broadcast below.

_____

_____

_____

_____

_____

_____

Lesson 11

Dust Storm!

As you read about the causes of the Dust Bowl, think of how people used and changed the land in this region.

In the 1860s many Americans in the East wanted land to farm. The government wanted land in the Midwest and Great Plains to be settled. It offered free land in this region to people who would farm it. Thousands of settlers accepted the offer. A westward move was on!

Many settlers found farming on the Great Plains hard. The soil was fertile, but farther west, farmers could not depend on rainfall levels from year to year. This area was part of what Major Stephen Long had called the "Great American Desert." Farmers had to irrigate the land to water their crops. They pumped water from the ground using the power of the wind that swept over the plains.

Soon wheat and other grains grew where cattle used to graze. In fact, the farmers pushed the cattle ranchers farther west. Cattle overgrazed the thin grass of the western Great Plains, eating so much grass that it could not grow back. Large areas of soil were left bare to the wind.

Dust Bowl
Great Plains States

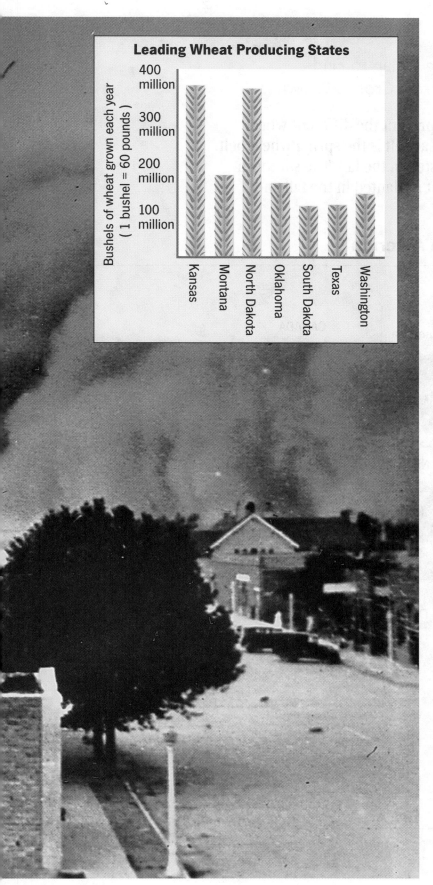

Leading Wheat Producing States

Bushels of wheat grown each year
(1 bushel = 60 pounds)

400 million
300 million
200 million
100 million

Kansas
Montana
North Dakota
Oklahoma
South Dakota
Texas
Washington

Then in the early 1900s, tractors and combines were invented. These new farm machines helped farmers plant even more land with wheat. But wheat did not hold the soil together against the wind the way grass roots had. When a long drought began in 1931, the soil dried up into fine, loose dust. And when the strong winds of the Great Plains began to blow, they picked up the dust and whirled it around.

One of the first huge dust storms carried tons of dirt from the Great Plains all the way to the East Coast! In cities like Chicago and Detroit the streetlights had to be turned on at noon because the blowing dust darkened the normally bright afternoon skies. Over the next four years of drought, dust storms swept across the plains. People and animals who were caught in the storms could barely breathe. Some became lost. Dust blew into homes so thickly that it had to be shoveled out. It buried cars and tractors, ruining them. Many people could not afford to keep their farms.

People tried to find solutions to help reduce the power of the wind. One of the solutions was to plant lines of trees to break the wind's force. Another solution included building wooden-slated wind fences. Farmers began using these methods to conserve, or protect, the soil from the wind. Dust storms still occur during dry periods on the plains, but soil conservation has reduced the negative effects.

The skies of the Great Plains were once blackened by mighty clouds of dust.

Lesson 11

MAP SKILLS
Using a Map to Find
Where a Crop Is Grown

The map below uses shades of color to represent the different wheat belts in North America. The northern wheat belt is the spring wheat belt. Wheat is planted in the spring and harvested in the fall. The southern wheat belt is the winter wheat belt. Wheat is planted in the fall and harvested in the spring.

Wheat-Producing Areas of North America

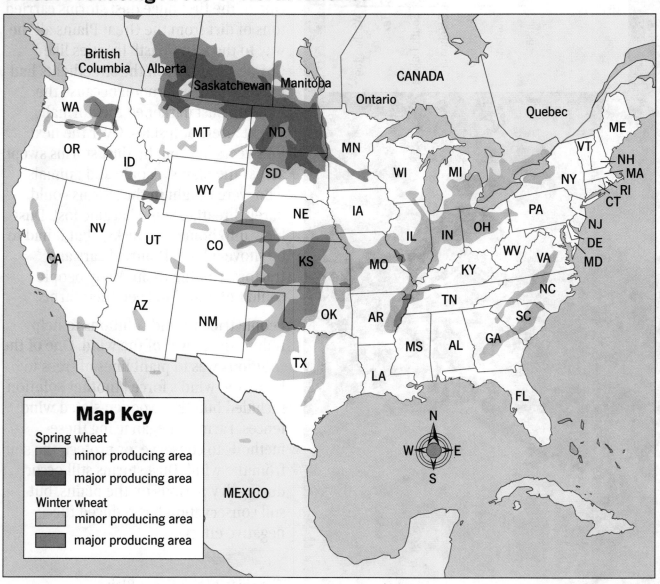

Map Key

Spring wheat
- minor producing area
- major producing area

Winter wheat
- minor producing area
- major producing area

A. Look at the map and complete the following.

1. Underline the names of the states in the northern spring wheat belt.
2. Circle the names of the states in the southern winter wheat belt.

B. Answer the following questions about where the different kinds of wheat are grown.

1. Which two states grow the most wheat? _____

_____

2. Which states produce both spring and winter wheat? _____

_____

_____

3. Which states are minor producers of winter wheat? _____

_____

_____

_____

_____

4. Is there more winter wheat or spring wheat grown in the United States? _____

_____

5. Which three spring wheat producing states grow the smallest amount of spring wheat?

_____

_____

_____

6. Where in North America is the most spring wheat grown?

_____

_____

Lesson 11

ACTIVITY
Find out about crops that are grown on the plains and how they are used.

Growing Grains

The **BIG** Geographic Question

What agricultural products do we get from the Midwestern plains?

In the article you read about how people started growing wheat on the plains and what effect it had on the land. The map skills lesson showed where wheat is grown in the Midwest today. Now find out more about crops grown in the Midwest and the uses people have for them.

A. Use the Almanac to find where wheat, corn, and soybeans are grown in the Midwest. List the crops and the Midwestern plains states where each crop is grown.

Crop	Where Grown
Wheat	
Corn	
Soybeans	

B. Find out how each crop is used. First look through your home for products made from each crop. Then find out about other uses. Try to find at least four uses for each crop and list them below. At least two uses should be something other than food items.

_____

_____

_____

_____

_____

C. On a separate sheet of paper, make a chart like the one below. List the crops you learned about. List the Midwestern states where they are grown, and picture or list some uses of each crop.

Crop	Where Grown	Uses
Wheat	North Dakota	bread

D. Choose one Midwestern plains crop. Write about how important this crop is in people's everyday lives. What might happen if flooding or storms destroyed much of this crop during a growing season?

_____

_____

_____

_____

_____

_____

_____

_____

_____

_____

_____

Lesson 12
A Southwestern Scrapbook

As you read about the American Southwest of long ago, think about how different cultures used similar environments and the land's natural resources in different ways.

The southwestern region of the United States is like a scrapbook. Pictures of its people, cultures, and communities give us many clues about the land long ago. Each culture used the natural resources of Earth: stone, clay, and mud. Each culture, however, used the resources in very different ways.

Long before the United States became a nation, prehistoric people such as the Anasazi tribe made their homes in the Southwest. They were known as the "cliff dwellers" because they had a unique way of building shelters using Earth's resources. They built their homes and communities into the sides of steep hills and mountains. The Zuni Native American tribe built apartmentlike villages. These Native Americans became known as the Pueblos, the Spanish word for "town" or "village." The pueblo style of architecture is still used today. The buildings appear to grow out of the land; some even look like landforms.

The largest early pueblo building was Pueblo Bonito in New Mexico, built between 900 and 1000 A.D. Pueblo Bonito has over 800 rooms.

In the 1500s the Spanish established towns across the Southwest. Spanish towns built around a central plaza became part of the architecture of the Southwest. They used adobe—bricks made from clay, water, straw, and grass. The word *adobe* is the Spanish name for sun-dried bricks or a house built with these special bricks. Adobe homes and other buildings stayed cool in summer and warm in winter.

In the 1800s another group of people moved to the Southwest. These people came from northern Europe and the eastern part of the United States and included many Germans. They used yet another resource to build their homes, forts, and other buildings—limestone. Limestone is a strong rock that can be carved and cut without splitting.

People traveling in the Southwest today can find traces of all of these cultures. They can see cliff dwellings, adobe pueblos, and limestone structures. Each one is a reminder of the people who lived in the Southwest many years ago.

The Alamo in San Antonio, Texas, is built of limestone. It was built in 1718 as a Catholic Mission and was later used by Texans as a military fort.

MAP SKILLS Using Maps to Understand Migration Patterns

Migration is the movement of people from one place to another. A population map shows the movement patterns in a specific region. It helps us to understand how and why people move and where they settle.

Migrant workers spend the winter months in California, Texas, and Florida. Then they move to work the harvest of crops farther north.

Major Routes of Migrant Workers in the United States

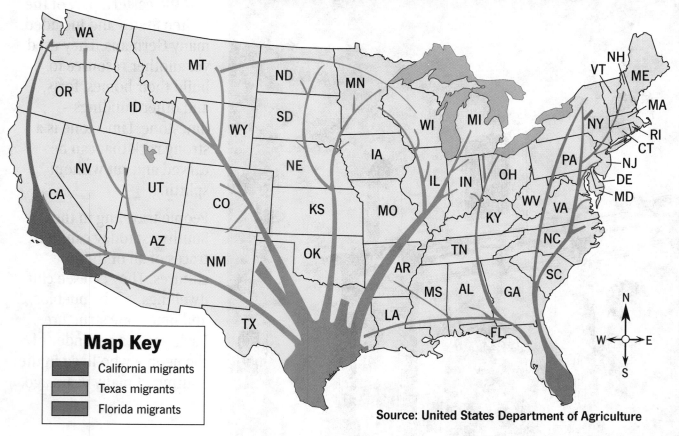

Map Key

- California migrants
- Texas migrants
- Florida migrants

Source: United States Department of Agriculture

A. Look at the migration map. Trace these three states on the map.

1. California (CA)

2. Texas (TX)

3. Florida (FL)

B. Use the map to complete the following.

1. What is the source of this migration map?

_____

2. Read the map's title. What do the three states represent on the map?

_____

_____

3. In which direction did the workers migrate?

_____

4. Why do you think the farm workers migrate to the north in the summer?

_____

_____

5. Suggest other reasons why people might move from one place to another.

_____

_____

6. Choose one migration route and circle it on your map. Describe how the route moves and changes. Be sure to use direction words (north, south, east, west) and the state names in your description.

_____

_____

_____

Lesson 12

ACTIVITY Create a cultural collage of the Southwest.

A Southwestern Collage

The **BIG** Geographic Question

Who are the people of the Southwest, and what are their customs?

From the article you learned about different cultures that lived in the Southwest long ago. The map skills lesson showed you how to use a map to understand movement patterns of migrant workers. Now make a collage of pictures and words about the cultures of the Southwest, past and present.

A. Write the names of the states in the southwestern region of the United States.

_____

_____

B. In the box below, draw a simple map of the Southwest. Label the states.

C. Use the article and the Almanac to find out more about the Southwest. As you learn about the region, take notes about the different people who live there and their cultures.

Characteristics of	Notes
People	
Food	
Language	
Clothing	
Customs	
Climate	
Land	
Water	
Buildings	

D. Look through magazines and newspapers. Cut out pictures or words that represent the southwestern cultures. You might want to draw pictures and write words of your own to add to your collection.

E. Put your drawings, pictures, and words together to make a large collage about the Southwest. Share your collage with the class.

Lesson 13

"A Grand" Canyon

As you read about the Grand Canyon, think about the effects of water on the land.

Imagine you're walking along a high desert plateau in the southwestern United States. Suddenly the ground seems to fall away from under your feet. A huge, gaping hole stretches for miles before you. It's so deep and wide that you can't see its bottom. You've found a **gorge**–a deep, narrow opening between steep and rocky walls or sides of mountains.

At the bottom of the Grand Canyon is the force that created it–the waters of the Colorado River. Over millions of years, this river has scoured, gouged, and carried away the land around it. It has carved a **canyon,** a deep valley with steep sides, usually with a stream running through it. The canyon is two billion years old!

The beautifully colored layers of rock in the canyon walls are like a history book. Each rock layer tells what the land was like at a certain time in the past. It also contains fossils that tell about the plants and animals that lived during that time.

The Grand Canyon is one of the United States' most popular places to visit. People marvel at its vastness and peacefulness.

74

Compared with the age of the canyon, humans have lived near it for only a short time—about 4,000 years. The earliest humans in the area were the Desert Culture people, a Native American tribe. Later the Anasazi dwelled there, from about 1,400 to 600 years ago. The Anasazi were probably descendants of the Desert Culture people, and the Navajo, Hopi, and Zuni may be descendants of the Anasazi. Anasazi cliff houses can still be seen in the canyon walls.

The first Europeans to see the canyon were Spanish explorers, who were led by Garcia Lopez de Cardenas in 1540. Three of his men tried to hike down to the river, but they failed. It was not until 1869 that a successful journey was made. An American scientist became the first person known to have led a group through the canyon. John Wesley Powell was a geologist who conducted geographical surveys of the Rocky Mountain region. His team traveled down the Colorado River in a small boat, surviving dangerous whitewater rapids. Powell recorded his daring voyage in pictures and words.

When President Theodore Roosevelt visited the canyon in 1906, he advised, "Keep it for your children and your children's children, and all who come after you, as the one great sight which every American should see." In 1919 his advice was taken, and the Grand Canyon was made a national park. Today four million people visit the canyon each year. Some adventurers brave the whitewaters of the river as Powell did more than 100 years ago.

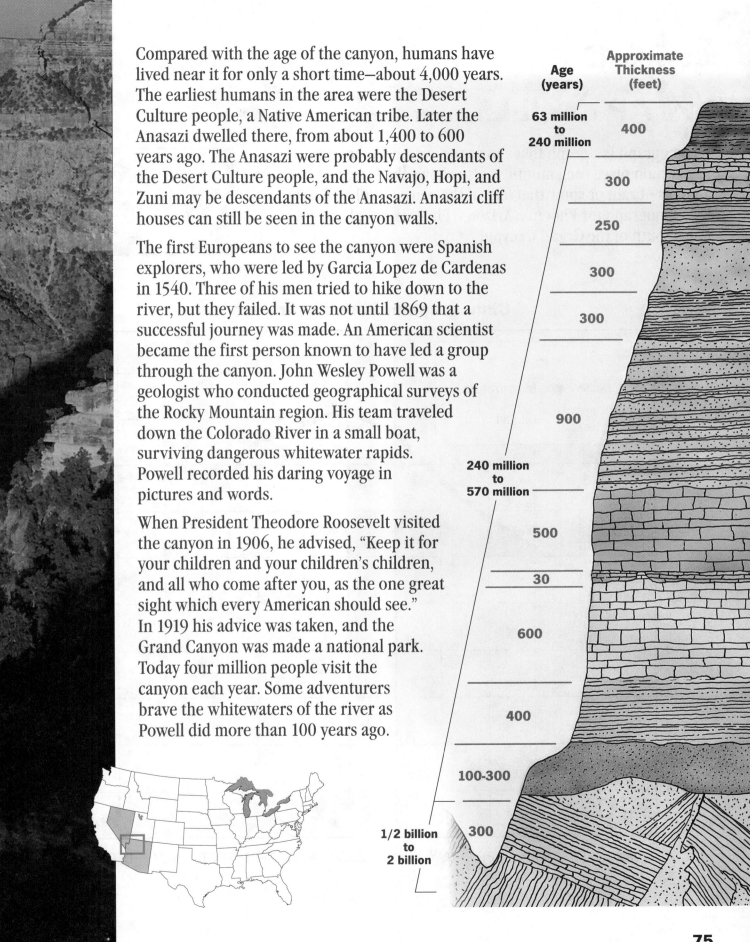

Age (years) | Approximate Thickness (feet)

63 million to 240 million — 400

300

250

300

300

900

240 million to 570 million

500

30

600

400

100-300

1/2 billion to 2 billion — 300

Lesson 13
MAP SKILLS
Using a Climograph to Find Temperatures and Rainfall

A climograph is a graph that shows the average temperature at a certain place each month. It also shows the average amount of rain or snow that falls at the place monthly. Below is a climograph for Phoenix, Arizona. Phoenix lies about 200 miles south of the Grand Canyon.

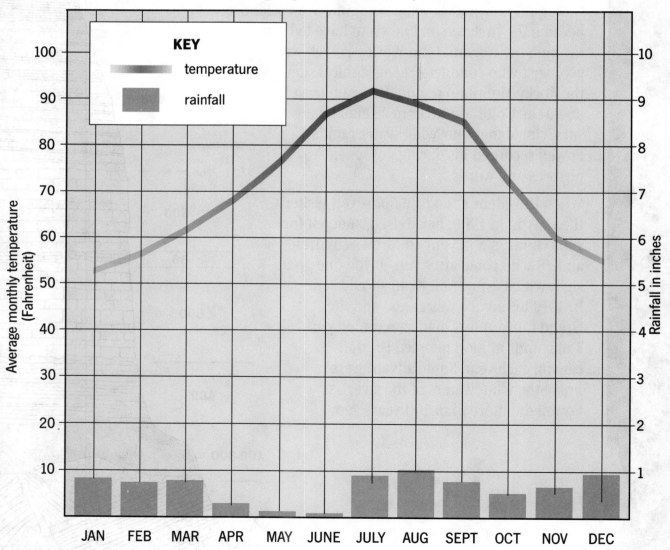

Climograph of Phoenix, Arizona

A. Look at the climograph and complete the following.

  1. Find and touch *January* at the bottom of the graph.

  2. Move your finger up until you touch the colorful, horizontal line on the graph.

  3. Move your finger to the left until you touch the column of numbers that stand for temperatures. The number nearest your finger is close to the average temperature in Phoenix in January. That temperature is about 52°F.

  4. Down the right side of the graph, circle the number that stands for inches of rainfall.

B. The bars on the graph show inches of rainfall in Phoenix. To find the average rainfall in Phoenix in January, follow these steps.

  1. Find and touch *January* at the bottom of the graph.

  2. Move your finger up until you touch the top of the bar.

  3. Move your finger to the right until you touch the column of inches of rainfall. The number nearest your finger is close to the average rainfall in Phoenix in January. That amount is almost 1 inch.

C. Answer the following questions using the climograph.

  1. What is the average temperature in Phoenix in July? _____

  2. What is the average rainfall in Phoenix in July? _____

D. Compare Phoenix's average temperature and rainfall in January and July. Would you say that Phoenix has a hot and dry climate, a hot and wet climate, a cool and wet climate, or a cool and dry climate?

January _____

July _____

E. Examine the temperatures and inches of rainfall for Phoenix for all twelve months. Write a brief statement explaining what you see.

_____

_____

Lesson 13

ACTIVITY
Find out the positive and negative effects of building dams in the Southwest.

Damming Wild Rivers

The **BIG**
Geographic Question

What are the positive and negative effects of building dams?

In the article you read about the Grand Canyon in the Southwest. In the map skills lesson you discovered how warm and dry states in the Southwest can be. Now study the efforts people have made to bring water to this area through the building of dams.

A. Dams are an important land feature found in many places. Put an **X** beside each statement below that describes how a dam might be used.

_____ To control the water flow of a river so that areas along the river get about the same amount of water

_____ To conserve water for dry periods that may occur throughout the year

_____ As a source of electrical power

B. Look at the map on page 125 of the Almanac and complete the chart below.

Dam	Location	River

C. Find out some reasons that these dams were built. List them on the left side of the chart below. Then find out some problems that have resulted from building these dams in a dry, semi-desert region. List them on the right side of the chart.

Building Dams

PROS (Good Effects)	CONS (Bad Effects)

D. Make a time line that shows the building of major dams in the Southwest. For each dam, mark the year it was completed and the river on which it was built.

1905	1915	1925	1935	1945	1955	1965	1975

E. Think about what you have discovered about dams. Write about the environmental effects—both good and bad—of building dams in the Southwest. Do you think dam building does more good than harm, or more harm than good? Tell what you think and why.

_____

_____

_____

_____

Lesson 14

High and Dry

As you read about the Mountain and Intermontane West region, think about the difficulty of crossing this rugged landscape.

The Mountain West region is a land of extremes. The region covers a huge area, but it has a small population. The region, which includes the states of Idaho, Montana, Wyoming, Nevada, Utah, and Colorado makes up almost one fourth of the land area of the United States. However, fewer than four of every 100 Americans live there.

One reason for the region's sparse population is the ruggedness of the land. Each state in the region has mountain ranges with peaks over two miles high. Many of these are in the Rocky Mountains. The Rockies are about 70 million years old. They rise as high as 14,433 feet above sea level at Mount Elbert in Colorado.

Some of North America's most important rivers start in the Rocky Mountains. The Colorado River and the Rio Grande begin in Colorado, and the Platte begins in Wyoming. The Missouri River starts in the Rocky Mountains of Montana. The direction in which these rivers flow is determined by the **continental divide,** an imaginary line running along the Rocky Mountains.

Rivers on the east side of the divide flow to the Atlantic Ocean by way of the Gulf of Mexico, while rivers on the west side flow to the Pacific Ocean. This means that raindrops that fall only a few feet apart in the Rockies may end up in different oceans thousands of miles apart.

Yellowstone Park in Wyoming has waterfalls, hot springs, and geysers. The most famous is "Old Faithful," which shoots up every 74 minutes.

The fact that the Mountain West is an arid, or dry, area despite its many rivers and lakes is another example of the extremes of this region. In fact, Nevada and Utah, part of the Intermontane West, are the two driest states in the United States. The Intermontane West is the area between the Rocky Mountains and the Pacific Coastal Range.

The Great Salt Lake of Utah is one of the natural wonders of the world. Although most lakes contain fresh water, this lake is saltier than ocean water. It is a lake with no outlets, so its water does not drain away. Instead, in the dry climate of the Intermontane West region the water **evaporates,** or changes to a mist, but leaves salt behind. The Bonneville Salt Flats near Salt Lake City, Utah, are an example of what happens when salt builds up and remains in the soil. This flat stretch of salt desert is a favorite spot for land speed trials.

Although the arid, rugged land and salty waters of the Mountain West region seem to have attracted few to live there, more and more people are visiting this land of extremes each year.

The spectacular sandstone pillars of Monument Valley in Utah have been the backdrop for many movies.

Lesson 14
MAP SKILLS
Comparing Maps to Learn About the History of an Area

A historical map shows information about a place's past. A comtemporary map shows information about the present. The maps below show routes used by people traveling west in the past and present.

1860s Overland Routes

Map Key

—— Mormon Trail

- - - Santa Fe Trail

- - - Oregon Trail

—— California Trail

• Cities

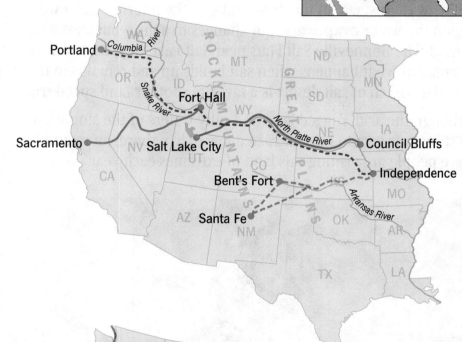

1990s Interstate Highway System

Map Key

—(5)— Interstate

• Cities

A. Study the 1860 map to find answers to these questions.

1. The routes of the Mormon Trail, the Oregon Trail, and the Sante Fe Trail each followed a river for part of the journey. Why do you think they did this?

_____

2. What major obstacle did each of the trails cross? _____

_____

3. The California Trail split from the Oregon Trail at Fort Hall. Which trail do you think was an easier route to follow?

_____

_____

B. Study the historical 1860s and contemporary 1990s maps to find answers to these questions.

1. People travel across the country a little differently today than they did in 1860. What does the 1990s map tell you about the West today? _____

_____

_____

2. Why would some routes on the contemporary map be the same as

the routes on the historical map? _____

_____

3. Which of the communities on the historical map are also on the

contemporary map? _____

_____

4. Which of the communities on the historical map are not on the

contemporary map? What does this tell you? _____

_____

_____

_____

Lesson 14

ACTIVITY Compare and contrast two urban centers of the West.

Two "Hot" Spots

The **BIG** Geographic Question

What physical and human features have attracted people to two very different desert communities?

In the article you read about the extremes of the Mountain and Intermontane West. The map skills lesson helped you compare past and present travel routes to the West. Now learn what has made two specific communities of the Intermontane West important urban centers.

A. In the Almanac read about the settling of Salt Lake City, Utah, and Las Vegas, Nevada. Use the chart below to help you organize your answers to the following questions.

Question	Salt Lake City, Utah	Las Vegas, Nevada
1. When was it settled?		
2. Who settled it?		
3. Why did they settle there?		

B. Use what you know about the physical features and climate of the region to help you answer these questions.

Question	Salt Lake City, Utah	Las Vegas, Nevada
1. What problems might settlers of this location face?		
2. How did the people of the region overcome the problems?		
3. Were these solutions successful?		

C. Read about the communities today. Think about how they have changed since they were first settled.

Question	Salt Lake City, Utah	Las Vegas, Nevada
1. What things are important to the city today?		
2. What role does tourism play in the community now?		

D. Write a brief article comparing and contrasting the two urban centers of Salt Lake City and Las Vegas. To go with your article, draw a picture to show the differences in the styles of the two communities.

Lesson 15

The Diversity of California

As you read about California, notice its many different physical and human features.

California is located in the southwest corner of the United States. It has the most **diverse,** or varied, physical and human features of any state in North America. On its eastern border are the rugged Sierra Nevada mountains. On its western border is the vast Pacific Ocean. In the north stands the active Lassen Peak volcano. The south is dominated by the arid Mojave desert. For many years, the desert and mountains almost cut California off from the rest of the country.

The people of California are as diverse as its land. Over time, the people living in the state have included Native Americans, Spanish explorers, missionaries, Chinese immigrants, and African Americans. Large numbers of Asians and Hispanics are more recent arrivals.

California's mild climate and varied natural scenery have long attracted people from all over the world. However, its Pacific coast location and vast population growth have not come without problems. Two major problems that California faces concern its natural resources and its natural landscape.

The Mojave Desert covers much of southern California.

Earthquakes, which are fairly common in California, can be devastating.

Los Angeles, a city with little water of its own, has a population of over three million. By building a system of aqueducts, or water channels, the city has been able to bring in enough water to keep growing. Water projects have also been important in California's agricultural industry. Much of California's farmland is too dry for growing crops. Irrigation brought water to these areas and made it possible for the state's farmers to lead the country in growing fruits and vegetables.

California has more than a dozen major **faults,** which are fractures or breaks in Earth's crust. The San Andreas Fault, which stretches nearly the length of California's Pacific coast, has been the site of several severe **earthquakes.** An earthquake is a sudden shaking or shock inside Earth that causes movement on its surface. San Francisco, located toward northern California, was hit hard by one in 1906, and by another in 1989. Los Angeles, located toward southern California, had severe earthquakes in 1971 and 1994. These cities were badly damaged but quickly rebuilt.

Nonetheless, the attractions of California are many and just as diverse as its land, people, and problems. For example, Central Valley, located between the Coast Ranges and the Sierra Nevada mountains, is a major agricultural region. Imperial Valley is another important agricultural area located at the southern tip of California, between San Diego and the Colorado River. In contrast, Silicon Valley, which stretches from south of San Francisco through Palo Alto, is where much of today's computer technology is produced. Also, the many movie studios in Hollywood have made Los Angeles the entertainment capital of the world.

California is a state of great wealth and diversity, in terms of land, products, people, and perils. People come from all over the world to California's exciting cities, further adding to the diversity of this region.

Lesson 15
MAP SKILLS Using Physical and Thematic Maps

One of California's biggest problems has been earthquakes that occur along the San Andreas Fault. A fault is a fracture or break in Earth's crust. The San Andreas Fault is one of the most active in the world and is very dangerous. While most faults lie deep beneath Earth's surface, some, like the San Andreas Fault, are visible. It looks like a gash in the land.

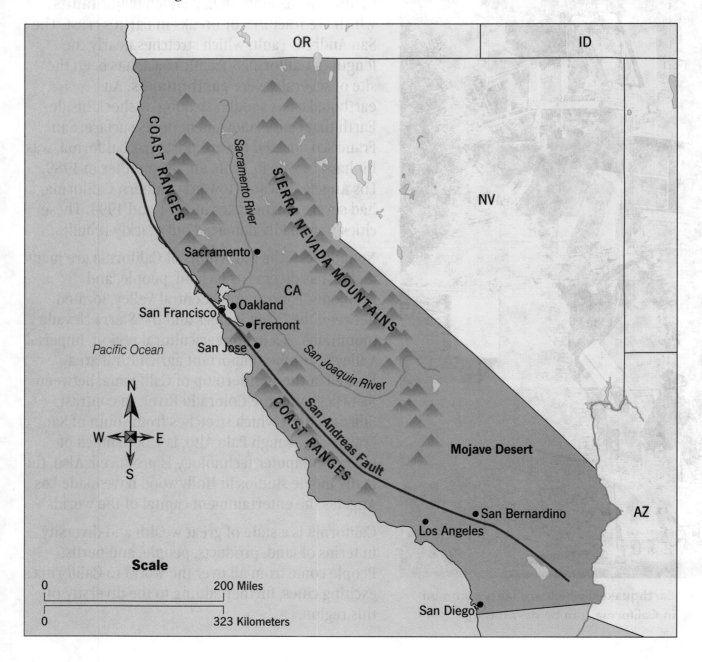

OR

ID

NV

CA

AZ

COAST RANGES

Sacramento River

SIERRA NEVADA MOUNTAINS

Sacramento

Oakland

San Francisco

Fremont

San Jose

Pacific Ocean

San Joaquin River

COAST RANGES

San Andreas Fault

Mojave Desert

San Bernardino

Los Angeles

San Diego

N
W   E
S

Scale

0   200 Miles

0   323 Kilometers

A. Study the physical map of California. Use the information on the map to complete the following.

1. Trace the line of the San Andreas Fault on the map.

2. How long is the San Andreas Fault? _____

3. What major cities are located on or near the fault? _____

_____

4. Write a brief desciption of where the fault is located, including where

it begins and ends. _____

_____

B. Continue to study the map, particularly the San Andreas Fault. Complete the following.

1. Which California city has been hit by two severe earthquakes, one

in 1906 and another in 1989? _____

_____

2. Use the chart below to describe the city, including where in California it is located and its land and water features.

City Name	
Where It Is Located	
Land and Water Features	

3. Do you think the city's location makes it more vulnerable to the activity of the San Andreas Fault than other cities located along the Fault? Why?

_____

4. How does the San Andreas Fault affect life in California? _____

_____

Lesson 15

ACTIVITY Illustrate the history of water management in California.

Where's the Water?

The **BIG** Geographic Question

What are some effects of water management on communities?

In the article you read about the diversity of California's human and physical features. The map skills lesson focused on the fragile San Andreas Fault in California. Now consider one of the state's other main concerns, the strained water supply. Create a model of California's physical relief to illustrate the location of the state's major efforts to reduce this strain.

A. California has four geographic regions. Find out about the regions. What are their water needs? Copy the following chart onto a sheet of paper. Complete the chart using the Almanac.

Region	Cities	Natural Features	Economic Activities	Source of Water
The Coast				
The Central Valley				
The Mountains				
The Deserts				

B. Use the Almanac to find out about these projects and complete the chart.

Project	Starting Point	Ending Point
Los Angeles Aqueduct		
Colorado River Aqueduct		
Hetch Hetchy Aqueduct		

C. Make your own relief model of California. Mix two parts salt (2 cups) to one part flour (1 cup), and some tap water (3/4 cup) to produce a moldable clay. Mix in a bowl and add 2–3 drops of cooking oil.

1. Plan your map on paper, marking the locations of California's four geographic regions.

2. Transfer the clay mixture to the map. Mold the mountains and other geographic features in relief and allow your model to dry.

3. When the model is completely dry, paint the major regions different colors using tempera or poster paints.

4. Use the information you have gathered from the charts to locate and illustrate the following features on your model:

   • Major urban areas

   • Major agricultural regions

   • Major water projects (dams, aqueducts, and reservoirs)

D. Think about the role water has played in the growth of California's population and industries. What problems do you think California might have with water in the future? What solutions would you recommend?

_____

_____

_____

_____

_____

Lesson 16

Timber!

As you read about the Pacific Northwest, consider what conditions promote the growth of the lush forests and what happens to them.

For hundreds of years, timber has been an important resource to people who live in the Pacific Northwest. Since the 1500s the people living there have depended on the wood from the forest to build and fuel their homes and to build canoes for traveling.

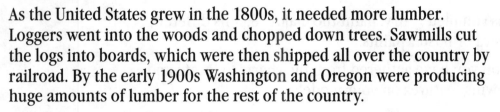

As the United States grew in the 1800s, it needed more lumber. Loggers went into the woods and chopped down trees. Sawmills cut the logs into boards, which were then shipped all over the country by railroad. By the early 1900s Washington and Oregon were producing huge amounts of lumber for the rest of the country.

Lush forests thrive in the Pacific Northwest states due to the large amount of rainfall they receive. The western side of Washington and Oregon receives up to 130 inches of rain per year in some places. The eastern side, however, is very dry. This difference is due to **orographic precipitation,** which occurs when a flow of warm air is forced upward by a rise in land, such as a hill or mountain.

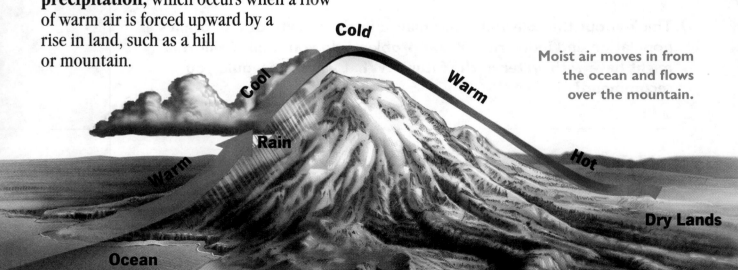

Cold

Cool

Warm

Moist air moves in from the ocean and flows over the mountain.

Rain

Warm

Hot

Ocean

Dry Lands

Four different systems for cutting timber

Clear-cut

Removes all trees in a large area

Seedcut

Leaves a few trees in an area to provide seeds for a new crop

Shelterwood cut

Leaves a few trees in an area for other trees that need shade to grow

Selective cut

Harvests mature trees to make room for new trees to grow

The lumber industry began to decline in the 1980s when fewer people in the United States built new homes. As a result, there was less demand for lumber. This problem became bigger in later years when wood was replaced by artificial products. Thousands of lumber workers lost their jobs. However, there was a need for lumber in Japan, a country with few timber resources of its own. Ships carried lumber across the Pacific Ocean to Japan, and many jobs were saved.

Another problem for the lumber industry centered around an argument over how to harvest the trees. This debate continues today. For many years, the most widespread method of logging has been **clear-cutting,** using giant sawing machines to strip entire hillsides of trees. The area is then reseeded with a single type of tree. The trees grow and are ready to harvest again in about 70 years.

Although it is efficient, clear-cutting causes **erosion,** the gradual wearing away of the earth. It leaves behind patches that are unsightly. Clear-cutting is also harmful to species of wildlife that inhabit the **old-growth forest,** or forest that has existed for a long time. Some of these animals, including the northern spotted owl, have almost become extinct as their habitat has been destroyed.

Timber in the Pacific Northwest continues to be an important resource. Unfortunately, the use of this resource has become an issue that requires people to weigh heavy consequences and make tough choices.

Lesson 16
MAP SKILLS Using Thematic Maps

In the Pacific Northwest warm, moist air moves in from the Pacific Ocean. As the air reaches the land, it flows up over the Coastal Range of mountains. It then flows down and across the Willamette Valley until it reaches a second, higher range of mountains, the Cascades. Look at the map below.

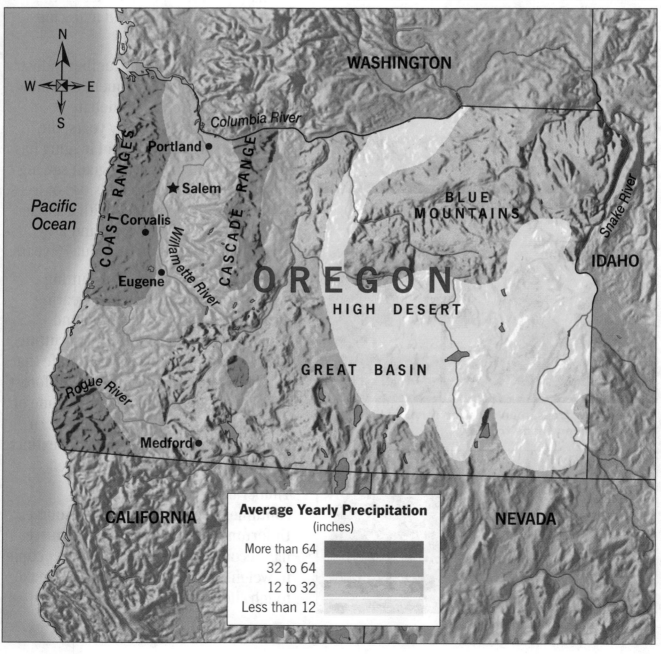

Average Yearly Precipitation
(inches)

More than 64
32 to 64
12 to 32
Less than 12

A. Study the map and the precipitation key.

1. What areas have the highest average precipitation? _____

2. What physical features are found in or near the areas that have the highest

precipitation? _____

3. What areas show the lowest average precipitation? _____

_____

4. What physical features are found in the areas that have the lowest

precipitation? _____

5. Look back at the diagram of orographic precipitation in the article on
page 92. Now look at the map of Oregon and the precipitation key.
Draw the diagram onto the map in the appropriate place.

**B. Look at the map diagram you have created and answer the
following questions.**

1. What happens to the air as it moves up the ocean side of the mountain?

_____

_____

2. Where does it rain? _____

_____

3. What is the air like as it moves down the far side of the mountain?

_____

_____

4. How does this precipitation affect the timber industry in the area?

_____

_____

Lesson 16

ACTIVITY Learn to see different sides of an environmental debate.

Depending on Forests

The **BIG**
Geographic Question

How can forest management issues be resolved?

From the article you learned about the vast amounts of timber in the Pacific Northwest and the benefits and problems connected with it. The map skills lesson showed you how the west side of Oregon gets more rain, creating forests on the western side of the state. Now investigate the debate over logging practices in the Pacific Northwest.

A. Answer the following questions about timber.

1. What are some items in your home or school that are made from timber?

_____

_____

2. Where do you think the timber to make these items comes from?

_____

B. Imagine that the items you listed above depended on timber from the Pacific Northwest.

1. How does rainfall affect the growth of trees in this region?

_____

_____

2. What is the most common method of harvesting the trees in the

Pacific Northwest? _____

C. Many people disagree over whether or not to harvest trees in the Pacific Northwest. What are some of the advantages of cutting down trees there? What are some of the disadvantages? Put your answers on the chart below.

Advantages	Disadvantages

D. Different groups of people feel strongly about the issue of cutting or not cutting down timber in the Pacific Northwest. Choose one of the following groups of people and think about how they would answer the questions below.

furniture makers

loggers

wildlife scientists

1. How does cutting down timber in the Pacific Northwest affect you?

_____

_____

2. What solutions would you suggest for the issue of cutting or not cutting down the timber?

_____

_____

E. Write a proposal to the United States Secretary of the Interior about this issue. State your view based upon the group that you chose. Take into account how others feel on the issue. What kinds of solutions are there that everyone could agree to?

Lesson 17
Island Treasures

As you read about the Hawaiian Islands, think about how its geography affects the lives of the people who live there.

Hawaii is the only state of our nation that is not located on the North American continent. It is a chain of volcanic islands miles away in the Pacific Ocean. The islands were created by volcanoes that have emerged from the ocean to form Hawaii's mountains.

The first people to live on Hawaii are thought to have traveled 2,500 miles in canoes from the Marquesas Islands, which are located southeast of Hawaii in the Pacific Ocean. About 800 years later more people came from Tahiti. These two Polynesian groups were the islands' native Hawaiians. They lived on the islands for several hundred years.

European explorers and foreigners from many lands began to stop at the Hawaiian Islands as a resting place on their journeys. The islands were seen as a crossroads between the eastern and western hemispheres. Foreigners saw Hawaii's lush vegetation, moderate climate, and ample rainfall as a good place for sugar plantations. When these plantations needed workers, Chinese people looking for jobs immigrated to Hawaii. Japanese and Filipino workers soon followed. Hawaii became a mixture of people from Asia, Europe, North America, and Africa.

The Hawaiian Islands have many active volcanoes.

The island of Oahu has nine of the state's ten largest cities and the highest population of any of the Hawaiian Islands. It was named Oahu, which means "gathering place," because in early times the chiefs from the different islands gathered there to meet. Today Hawaii's government meets in the capital city of Honolulu on Oahu. The gathering of the many cultures from East and West can still be seen in Oahu's population today.

Native foods such as *poi,* a starchy food from the stem of the taro plant, and *laulau,* spinach-like chopped taro leaves with fish and pork wrapped in plant leaves, are still very popular. Rice and fruits and vegetables grown on the islands are a major part of the Hawaiian diet. Foods introduced by immigrants reflect the cultural backgrounds of the varied population. Asian, European, and mainland American foods are featured in local Hawaiian supermarkets and restaurants today.

The Hawaiian Islands are best known for their location and climate. But the many different people who have inhabited its lands have influenced these treasured islands.

Pineapples are one of the main crops grown in Hawaii.

Lesson 17
MAP SKILLS Using Maps to Learn About Population

A **census** is a count of the people living in the United States. It is done every ten years. **Population density** means the concentration of people living in a particular area. The map below uses colors to show the population density of Hawaii according to the 1990 census.

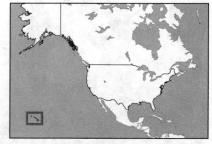

Population Density of the Hawaiian Islands

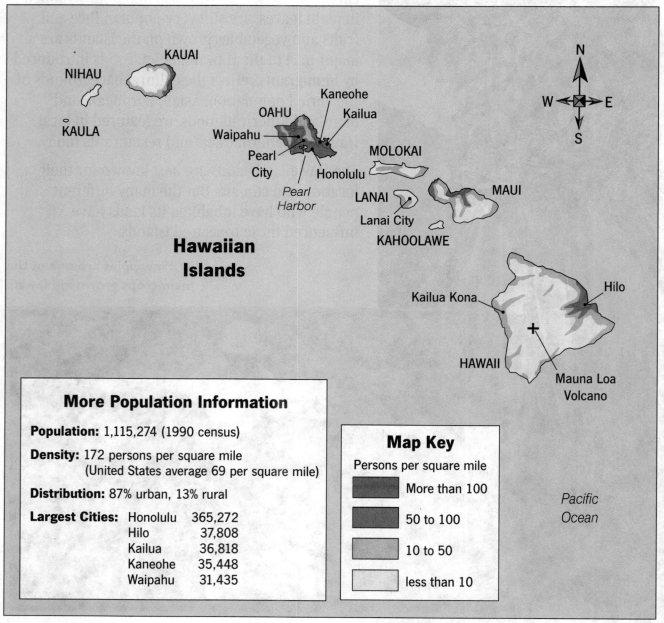

KAUAI

NIHAU

KAULA

Kaneohe

OAHU

Kailua

Waipahu

Pearl City

Pearl Harbor

Honolulu

MOLOKAI

LANAI

Lanai City

MAUI

KAHOOLAWE

Hawaiian Islands

Kailua Kona

Hilo

HAWAII

Mauna Loa Volcano

N W E S

More Population Information

Population: 1,115,274 (1990 census)

Density: 172 persons per square mile
(United States average 69 per square mile)

Distribution: 87% urban, 13% rural

Largest Cities:

Honolulu	365,272
Hilo	37,808
Kailua	36,818
Kaneohe	35,448
Waipahu	31,435

Map Key
Persons per square mile

- More than 100
- 50 to 100
- 10 to 50
- less than 10

Pacific Ocean

A. Look at the map and answer the questions below.

1. List the nine islands that make up the state of Hawaii.

   a. _____   f. _____

   b. _____   g. _____

   c. _____   h. _____

   d. _____   i. _____

   e. _____

2. Which color on the map represents the highest density, or concentration of people?

   _____

3. Which color represents the sparsely populated areas—areas with the fewest people?

   _____

B. *Urban* means "in, of, or like a city." *Rural* means "in, of, or like the country." Use the map to answer the following questions.

1. Which map color would you say best represents urban areas? _____

2. Which color would you say best represents rural areas? _____

3. How many people per square mile live on the northern tip of the

   island of Hawaii? _____

4. How do the coasts of the islands compare to the inland areas in population?

   _____

5. Which Hawaiian island has the highest population? _____

6. On which island are most of the cities located? _____

7. Compare the city locations with the population density and note your

   observations. _____

   _____

Lesson 17

ACTIVITY Discover how people have changed Hawaii's ecosystem.

People Causing Change

The BIG Geographic Question **How did people change the plant and animal life on Hawaii?**

From the article you learned that people from many cultures settled on the Hawaiian Islands. The map skills lesson showed the population densities of the islands. Now find out how people have affected the landscape of Hawaii.

A. **The ecosystem of Hawaii includes many living things that are not found anywhere else on Earth. Use the Almanac to find information about the unique life-forms of Hawaii, past and present, and about the effects people have had on the ecosystem.**

1. List some of Hawaii's unique plants and animals below. Circle any that

   are endangered. _____

   _____

   _____

   _____

2. Why do you think the ecosystem of Hawaii is unique? _____

   _____

   _____

   _____

   _____

B. Use what you have learned to complete the chart below. Tell how the climate affects each feature and how that feature affects the ecosystem. Add other features you have learned about.

Economic Activity	How Affected by Climate	Effect on Hawaii's Ecosystem
sugarcane plantations		
pineapple plantations		

C. Use the Almanac to find out where pineapples and sugarcane are grown in Hawaii. Then trace or draw a map of Hawaii and place symbols on the map to show the locations of the pineapples and sugarcane. Remember to draw a map key for your symbols.

D. Write a few sentences explaining the effects of human actions on Hawaii's ecosystem.

_____

_____

_____

_____

_____

_____

_____

103

Lesson 18

Journeying to Alaska

As you read about the geography of Alaska, think about how important it was to develop new forms of transportation and communication.

Alaska is so far north that almost one third of the state lies north of the Arctic Circle. Alaska has a very cold climate and two major mountain ranges. These conditions make it difficult to construct roads and highways there.

The Alaska Highway was built by the United States government during World War II to provide a land route for military equipment. The Dalton Highway was built in the early 1980s. It followed the route of the Trans-Alaska Pipeline. The pipeline was built to carry oil from Prudhoe Bay in northern Alaska to the port of Valdez in the south.

Other than on highways, land travel involves using dogsleds and ATVs (all-terrain vehicles). Dogsleds are one of the earliest means of transportation used by the native Alaskans—the Aleut and Inuit peoples. Today ATVs and snowmobiles are becoming more popular.

Highways have made a difference in Alaska, but airplanes have made an even greater impact. The airplane was introduced to Alaska around 1920. Airplanes brought news and supplies to areas that had little or no connection to the rest of the world. Smaller planes that land on water and planes with special protection from the harsh weather can reach isolated areas. Pilots who fly small airplanes to remote areas are known as bush pilots. They fly to every corner of the state.

Bush pilots often have no runways, so they land on glaciers or lakes. The people in these remote areas appreciate the service these pilots provide in transporting mail, people, and supplies.

Trans-Alaska Pipeline

Another means of transportation in Alaska is the state-owned Alaskan Railroad. It is used to transport materials and manufactured goods from the port cities of Seward and Whittier to Fairbanks and Anchorage, two of the state's largest cities. The Alaskan Railroad also operates passenger trains.

The building of highways in Alaska has connected cities and people to each other. The developments in transportation have made travel and communication easier for people in even the most remote of regions.

Some Alaskans still rely on dogsleds as transportation.

As transportation and communication were developed, more people came to Alaska. The Aleut and Inuit began to realize that they were losing control of their homeland. In 1966 these groups came together to form the Alaskan Federation of Natives. They tried to reclaim the 280 million acres of land that belonged to their ancestors. The Alaska Native Claims Settlement Act of 1971 gave back to the native people 44 million acres of land and nearly $1 billion.

Lesson 18
MAP SKILLS Identifying Various Types of Lines on a Map

Maps that show boundary lines are called **political maps.**
Political maps can show cities, towns, highways, and
sometimes other features such as ferry routes, agricultural
areas, and military bases. Political maps can also show
international boundaries between different countries.

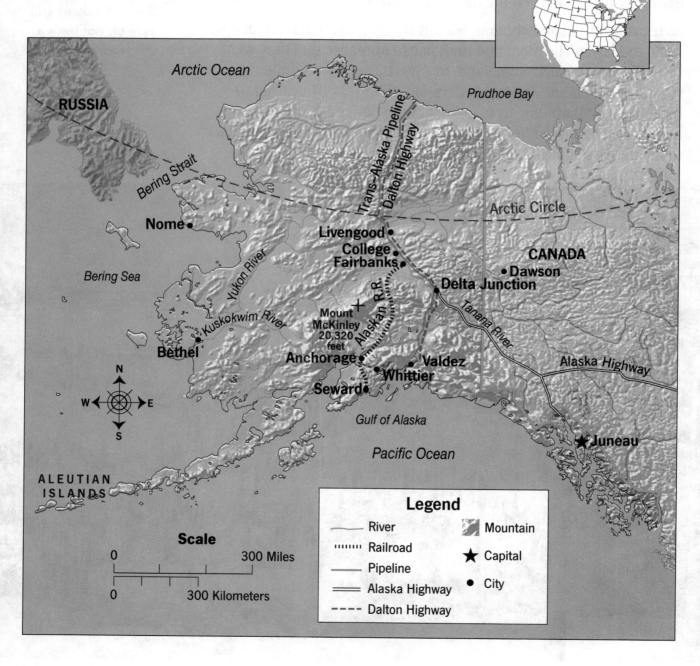

Legend

——— River

·········· Railroad

——— Pipeline

===== Alaska Highway

- - - - Dalton Highway

Mountain

★ Capital

● City

Scale

0   300 Miles

0   300 Kilometers

A. Look at the map and identify the following features.

1. Find an international boundary line and trace it with a red line.

2. Find the highway line and trace it with a yellow line.

3. Find the railroad line and trace it with a blue line.

4. Find the Trans-Alaska Pipeline and trace it with a green line.

B. Look at the legend on the map. Find these symbols and write what each one stands for.

1. """""" _____

2. —— _____

3. - - - - _____

4. ⌒ _____

C. Use the map to answer these questions about international boundaries.

1. What country borders the east side of Alaska? _____

2. How can you tell the boundary line between Alaska and that country? _____

_____

3. What country borders the west side of Alaska?_____

4. What body of water forms a boundary between Alaska and that country? _____

_____

D. Think about the boundaries between Alaska and Canada and Alaska and Russia. Write a few sentences explaining which is a physical boundary and which is a human boundary. Explain your answer.

_____

_____

_____

Lesson 18

ACTIVITY
Find out how communication and transportation can affect people.

Saving the Land

The **BIG** Geographic Question | How have the lives of native Alaskans been affected by transportation and communication developments?

From the article you learned that as more ways of traveling to and in Alaska came about, more people went there. The map skills lesson showed how different lines on a map represent features of Alaska. Now find out what happened with the land in Alaska.

A. Use the information in the article, map skills lesson, and Almanac to complete the following about Alaska.

1. Describe the land of Alaska?

_____

_____

2. Who are the native people of Alaska?

_____

_____

3. How do the native people usually travel in Alaska?

_____

_____

4. How has the land of Alaska been used to connect it to other places?

_____

_____

_____

B. Review the information you learned about why the native Alaskans wanted to reclaim land and how the United States government responded to their claims. Organize the information you find on the chart below.

People	Actions Regarding Land in Alaska
Native Alaskans' claims	
United States government's response	

C. Use what you have learned from the activity and the Almanac to role-play with a partner how you might have decided to divide the land and cash. Then write a short paragraph below describing your decision. Include a description of how your decision compares to what the Native Alaskans decided to do.

_____

_____

_____

_____

_____

_____

_____

_____

_____

_____

_____

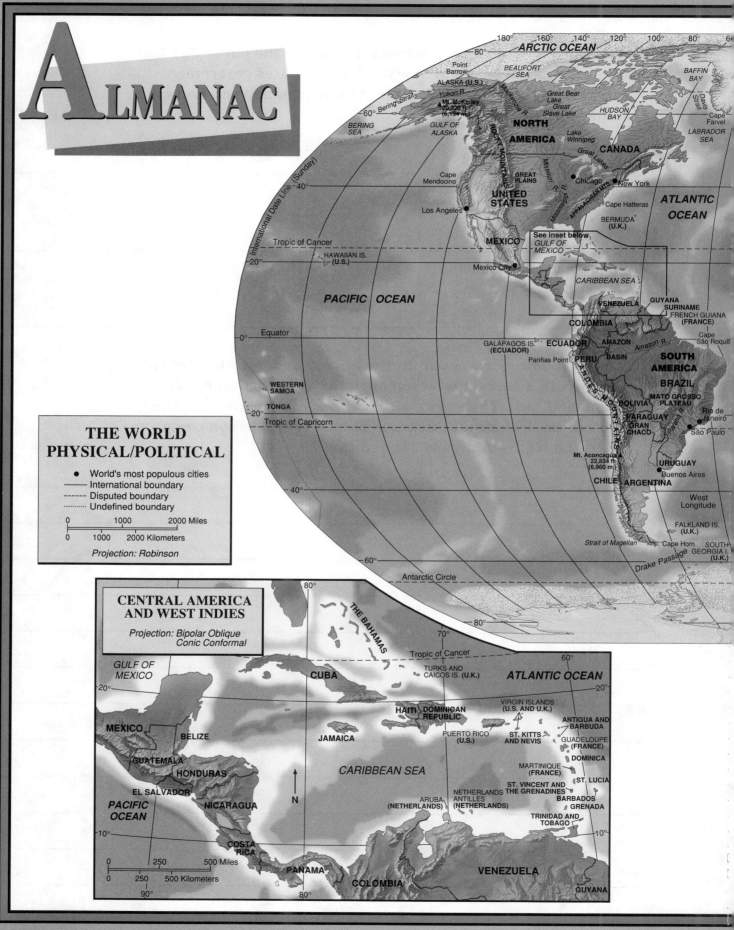

THE WORLD
PHYSICAL/POLITICAL

- • World's most populous cities
- —— International boundary
- ----- Disputed boundary
- .... Undefined boundary

0	1000	2000 Miles
0	1000	2000 Kilometers

Projection: Robinson

CENTRAL AMERICA
AND WEST INDIES

*Projection: Bipolar Oblique
Conic Conformal*

0	250	500 Miles
0	250	500 Kilometers

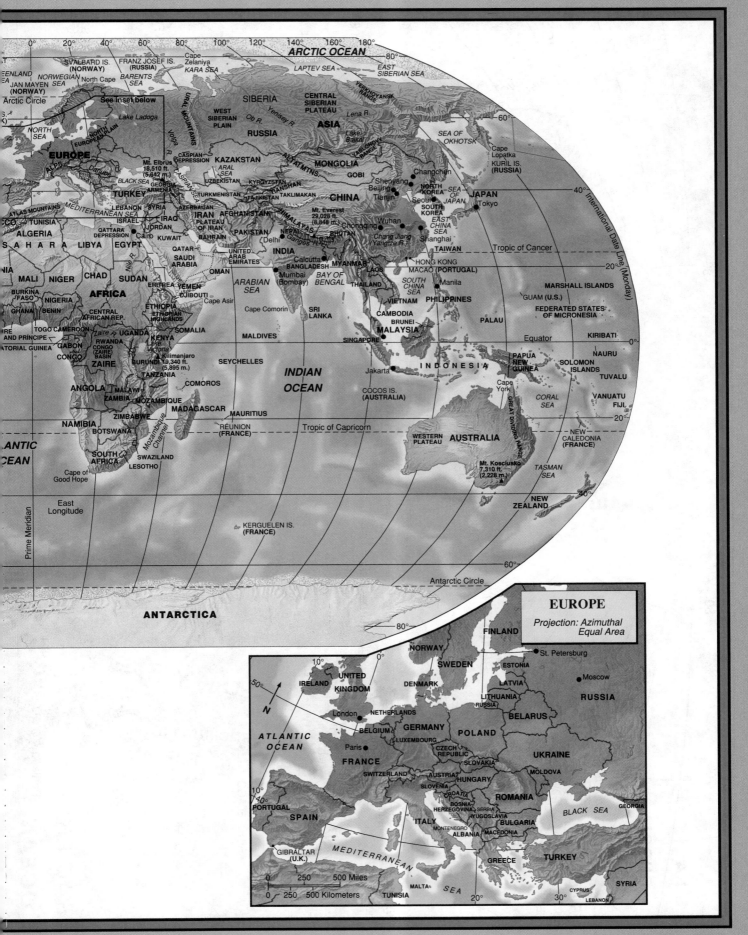

ARCTIC OCEAN

0° 20° 40° 60° 80° 100° 120° 140° 160° 180° 80°

Cape Zelaniya
SVALBARD IS. (NORWAY) FRANZ JOSEF IS. (RUSSIA) KARA SEA LAPTEV SEA EAST SIBERIAN SEA
NORWEGIAN SEA BARENTS SEA North Cape VERKHOYANSK RANGE
JAN MAYEN (NORWAY)
Arctic Circle
See Inset below
Lake Ladoga SIBERIA WEST SIBERIAN PLAIN CENTRAL SIBERIAN PLATEAU Lena R. 60°
NORTH SEA EUROPE Volga R. URAL MOUNTAINS Ob R. Yenisey R. ASIA Lake Baikal SEA OF OKHOTSK
ALPS Danube RUSSIA YABLONOVY RANGE Cape Lopatka
Mt. Elbrus 18,510 ft. (5,642 m.) CASPIAN DEPRESSION ARAL SEA KAZAKSTAN KURIL IS. (RUSSIA)
BLACK SEA GEORGIA CASPIAN SEA MONGOLIA Changchun 40°
TURKEY ARMENIA UZBEKISTAN KYRGYZSTAN TIAN SHAN ALTAI MTNS. Shenyang NORTH KOREA SEA OF JAPAN JAPAN
MEDITERRANEAN SEA LEBANON SYRIA AZERBAIJAN TURKMENISTAN TAJIKISTAN TAKLIMAKAN GOBI Beijing Tianjin Seoul SOUTH KOREA Tokyo
ATLAS MOUNTAINS IRAQ IRAN AFGHANISTAN HIMALAYAS CHINA Wuhan
TUNISIA ISRAEL JORDAN PLATEAU OF IRAN PAKISTAN Mt. Everest 29,028 ft. (8,848 m.) Chongqing Shanghai EAST CHINA SEA
ALGERIA LIBYA EGYPT QATTARA DEPRESSION Cairo KUWAIT BAHRAIN NEPAL BHUTAN Chang Jiang (Yangtze R.)
SAHARA Nile R. QATAR SAUDI ARABIA UNITED ARAB EMIRATES Delhi Ganges R. INDIA Calcutta MYANMAR TAIWAN Tropic of Cancer 20°
MALI NIGER CHAD SUDAN ERITREA YEMEN OMAN ARABIAN SEA Mumbai (Bombay) BANGLADESH BAY OF BENGAL LAOS HONG KONG MACAO (PORTUGAL)
BURKINA FASO ETHIOPIA DJIBOUTI THAILAND SOUTH CHINA SEA Manila MARSHALL ISLANDS
NIGERIA AFRICA Cape Asir Cape Comorin SRI LANKA VIETNAM PHILIPPINES GUAM (U.S.)
GHANA BENIN CENTRAL AFRICAN REP. ETHIOPIAN HIGHLANDS CAMBODIA FEDERATED STATES OF MICRONESIA
TOGO CAMEROON SOMALIA MALDIVES BRUNEI PALAU KIRIBATI
AND PRINCIPE GABON UGANDA KENYA Lake Victoria SEYCHELLES SINGAPORE MALAYSIA Equator 0°
EQUATORIAL GUINEA CONGO (ZAIRE) BASIN RWANDA BURUNDI Kilimanjaro 19,340 ft. (5,895 m.) INDIAN OCEAN INDONESIA NAURU
CONGO ZAIRE TANZANIA Jakarta PAPUA NEW GUINEA SOLOMON ISLANDS TUVALU
ANGOLA MALAWI COMOROS COCOS IS. (AUSTRALIA) Cape York VANUATU
ATLANTIC OCEAN NAMIBIA ZAMBIA MOZAMBIQUE MADAGASCAR MAURITIUS GREAT DIVIDING RANGE CORAL SEA NEW CALEDONIA (FRANCE) FIJI
BOTSWANA ZIMBABWE Mozambique Channel RÉUNION (FRANCE) Tropic of Capricorn WESTERN PLATEAU AUSTRALIA 20°
SOUTH AFRICA SWAZILAND Mt. Kosciusko 7,310 ft. (2,228 m.) TASMAN SEA
Cape of Good Hope LESOTHO NEW ZEALAND 30°
East Longitude KERGUELEN IS. (FRANCE)
Prime Meridian
60°
Antarctic Circle
ANTARCTICA 80°

EUROPE
Projection: Azimuthal Equal Area

FINLAND
NORWAY SWEDEN St. Petersburg
10° ESTONIA Moscow
50° IRELAND UNITED KINGDOM DENMARK LATVIA RUSSIA
N London NETHERLANDS LITHUANIA RUSSIA BELARUS
ATLANTIC OCEAN BELGIUM GERMANY POLAND
Paris LUXEMBOURG CZECH REPUBLIC UKRAINE
10° FRANCE SWITZERLAND SLOVAKIA MOLDOVA
40° AUSTRIA HUNGARY
PORTUGAL SPAIN SLOVENIA CROATIA ROMANIA
ITALY BOSNIA HERZEGOVINA SERBIA BLACK SEA GEORGIA
MONTENEGRO YUGOSLAVIA BULGARIA
GIBRALTAR (U.K.) ALBANIA MACEDONIA
250 500 Miles MEDITERRANEAN GREECE TURKEY
0 250 500 Kilometers MALTA SEA CYPRUS SYRIA
TUNISIA 20° 30° LEBANON

Arctic Circle

Denmark Strait

70°

Cape F

NEWF

GULF OF
ST. LAWRENCE (F

ST
Halifax

LABRADOR
SEA

KALAALLIT NUNAAT
(GREENLAND)
(DENMARK)

LABRADOR

L
A
B
R
A
D
O
R

Davis Strait

80°

BAFFIN
BAY

BAFFIN

ISLAND

Hudson Strait

UNGAVA
PEN.

Smallwood
Res.

S
H
I
E
L
D

Quebec

Lake
St. Lawrence

Montreal
Ottawa

R

Nares Str.

ELLESMERE
ISLAND

North
+
Pole

HUDSON
BAY

C
A
N
A
D
I
A
N

Lake
Superior

QUEEN ELIZABETH
ISLANDS

VICTORIA

ISLAND

Churchill

Nelson R.

Lake
Winnipeg

Winnipeg

ARCTIC OCEAN

BEAUFORT
SEA

Great Bear
Lake

Great Slave
Lake

CANADA

Lake
Athabasca

Churchill R.

Reindeer
Lake

Saskatchewan R.

Regina

Lake
Manitoba

GREAT

UNITED STATES

80°

Point
Barrow

Mackenzie River

Peace R.

Edmonton

Athabasca R.

North

South

Calgary

M

70°

CHUKCHI
SEA

MACKENZIE
MOUNTAINS

Mt. Logan
19,850 ft.
(6,050 m.)

Whitehorse

R O C K Y

Fraser R.

Columbia R.

Spokane

Snake

COLUM
PLATE

CASCADE RANGE

RUSSIA

Fairbanks

Yukon River

ALASKA (U.S.)

ALASKA RANGE

COAST MOUNTAINS

Vancouver

Victoria

Seattle

Portland

S I A

Bering Strait

RENCE I.

SEWARD
PEN.

Mt. McKinley
20,320 ft.
(6,194 m.)

Anchorage

Juneau

GULF OF
ALASKA

ALEXANDER
ARCHIPELAGO

QUEEN
CHARLOTTE
ISLANDS

VANCOUVER
ISLAND

PACIFIC

OCEAN

RING
SEA

PENINSULA

K I.

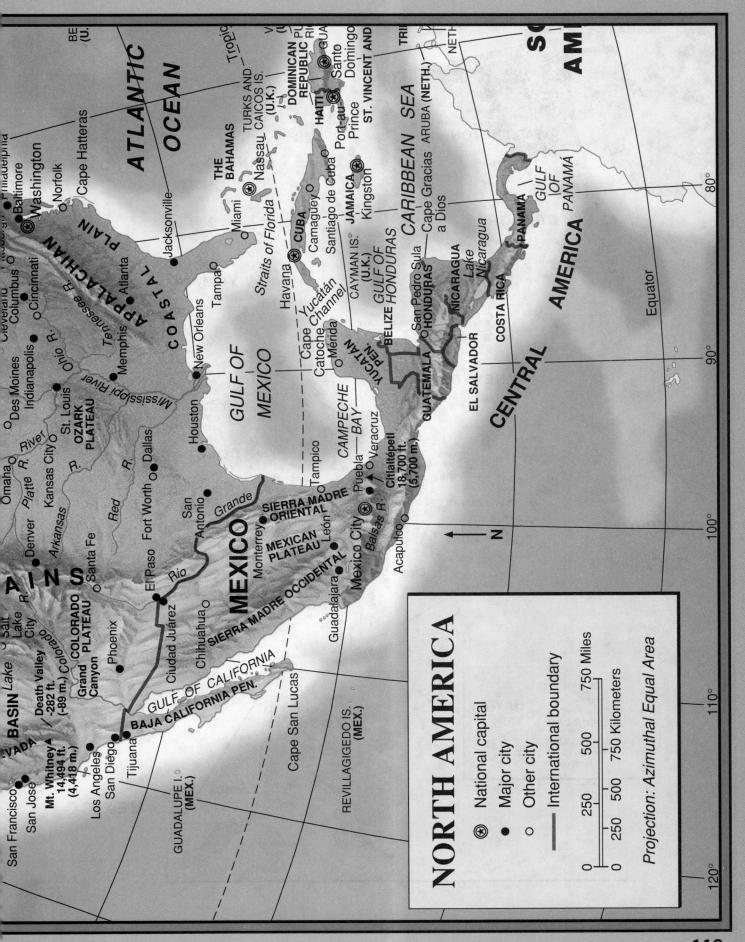

ATLANTIC OCEAN

Philadelphia

BE.
(U.

Baltimore
Washington

Cape Hatteras

Norfolk

Tropic

DOMINICAN
REPUBLIC

Santo
Domingo

TURKS AND
CAICOS IS.
(U.K.)

THE
BAHAMAS

Nassau

Port-au-
Prince

ST. VINCENT AND

HAITI

PU
RI

GUA

Jacksonville

Miami

Straits of Florida

CUBA

Camaguey

Santiago de Cuba

JAMAICA

Kingston

CARIBBEAN SEA

ARUBA (NETH.)

CAYMAN IS.
(U.K.)

Cape Gracias
a Dios

PANAMA

GULF
OF
PANAMA

TRI

NETH.

Cleveland

Columbus
Cincinnati

Atlanta

Atlantic

COASTAL
PLAIN

APPALACHIAN

Tampa

New Orleans

GULF OF
MEXICO

Cape
Catoche

Yucatán
Channel

GULF OF
HONDURAS

BELIZE

San Pedro Sula

HONDURAS

NICARAGUA

Lake
Nicaragua

COSTA RICA

CENTRAL AMERICA

Equator

80°

90°

Des Moines

Indianapolis

Memphis

Tennessee R.

Ohio R.

St. Louis

OZARK
PLATEAU

Mississippi River

Houston

Havana

Mérida

YUCATÁN
PEN.

CAMPECHE
BAY

GUATEMALA

EL SALVADOR

Omaha

Platte R.

Kansas City

Denver

Arkansas R.

Red R.

Santa Fe

Dallas

Fort Worth

San
Antonio

Grande

SIERRA MADRE
ORIENTAL

Monterrey

Tampico

Veracruz

Puebla

Citlaltépetl
18,700 ft.
(5,700 m.)

MEXICAN
PLATEAU

MEXICO

León

Mexico City

Balsas R.

Acapulco

N

100°

BASIN

VADA

Death Valley
/ -282 ft.
(-89 m.)

Colorado R.

Grand
Canyon

COLORADO
PLATEAU

Phoenix

Salt
Lake

Lake
City

El Paso

Ciudad Juárez

Rio

Chihuahua

SIERRA MADRE OCCIDENTAL

Guadalajara

AINS

R.

Mt. Whitney
14,494 ft.
(4,418 m.)

San
Diego

Tijuana

GULF OF CALIFORNIA

BAJA CALIFORNIA PEN.

Cape San Lucas

GUADALUPE I.
(MEX.)

REVILLAGIGEDO IS.
(MEX.)

San Francisco

San Jose

Los Angeles

110°

120°

NORTH AMERICA

⊛ National capital

● Major city

○ Other city

—— International boundary

| 0 | 250 | 500 | 750 Miles |
| 0 | 250 | 500 | 750 Kilometers |

Projection: Azimuthal Equal Area

SO
AM

CARIBBEAN SEA

113

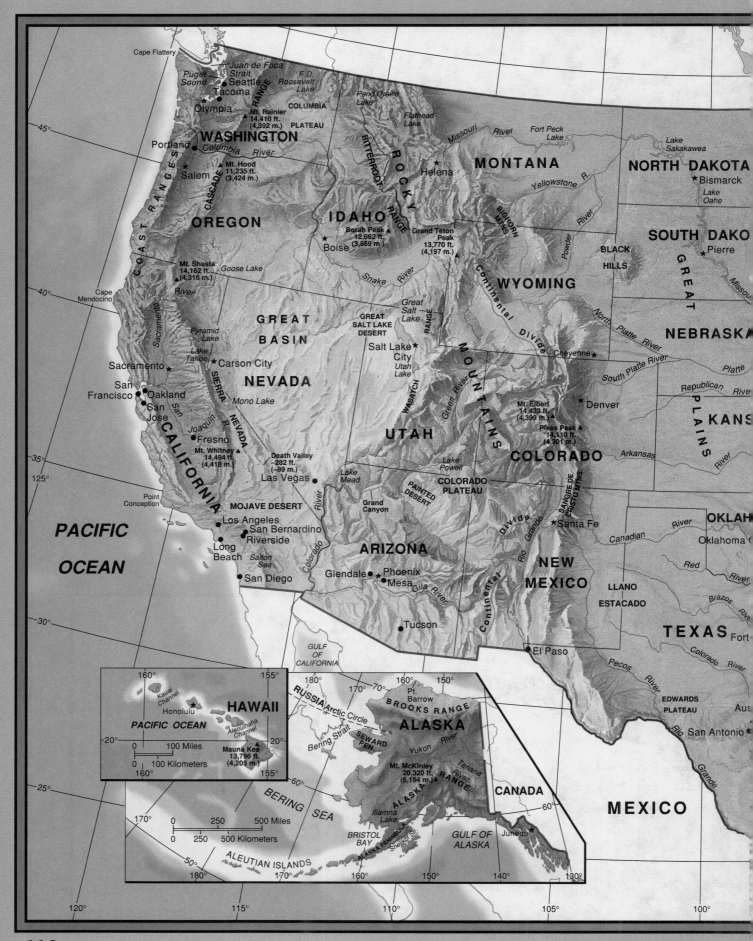

Cape Flattery
Juan de Fuca Strait
Puget Sound
Seattle
Tacoma
Olympia
Mt. Rainier 14,410 ft. (4,392 m.)
WASHINGTON
Portland
Columbia River
Salem
Mt. Hood 11,235 ft. (3,424 m.)
CASCADE RANGE
OREGON
Mt. Shasta 14,162 ft. (4,316 m.)
Goose Lake
River
Cape Mendocino
COAST RANGES
Sacramento
Pyramid Lake
Lake Tahoe
Sacramento
San Francisco
Oakland
San Jose
CALIFORNIA
Joaquin R.
SIERRA NEVADA
Fresno
Mt. Whitney 14,494 ft. (4,418 m.)
Point Conception
MOJAVE DESERT
Los Angeles
San Bernardino
Riverside
Long Beach
Salton Sea
San Diego

F.D. Roosevelt Lake
COLUMBIA PLATEAU
Pend Oreille Lake
Flathead Lake
BITTERROOT RANGE
ROCKY
Boise
Borah Peak 12,662 ft. (3,859 m.)
IDAHO
Snake River
GREAT BASIN
GREAT SALT LAKE DESERT
Great Salt Lake
Salt Lake City
Utah Lake
NEVADA
Carson City
Mono Lake
WASATCH
RANGE
Death Valley -282 ft. (-89 m.)
Las Vegas
Lake Mead
UTAH
COLORADO PLATEAU
Colorado River
Grand Canyon
PAINTED DESERT
ARIZONA
Glendale
Phoenix
Mesa
Gila River
Tucson

Helena
MONTANA
Missouri River
Fort Peck Lake
Yellowstone R.
BIGHORN MTNS.
Grand Teton Peak 13,770 ft. (4,197 m.)
Continental Divide
WYOMING
Powder River
North Platte River
Cheyenne
South Platte River
Green River
Mt. Elbert 14,433 ft. (4,399 m.)
Denver
Pikes Peak 14,110 ft. (4,301 m.)
COLORADO
Lake Powell
COLORADO PLATEAU
SANGRE DE CRISTO MTNS.
Continental Divide
Rio Grande
Santa Fe
NEW MEXICO
LLANO ESTACADO
El Paso

NORTH DAKOTA
Bismarck
Lake Sakakawea
Lake Oahe
SOUTH DAKOTA
Pierre
BLACK HILLS
GREAT
NEBRASKA
Platte
Republican River
KANSAS
Arkansas River
PLAINS
Missouri
OKLAHOMA
Oklahoma City
Canadian River
Red River
Brazos
TEXAS
EDWARDS PLATEAU
San Antonio
Pecos River
Colorado River
Rio Grande
MEXICO

PACIFIC OCEAN

Kauai Channel
Honolulu
HAWAII
PACIFIC OCEAN
0   100 Miles
0   100 Kilometers
Alenuihaha Channel
Mauna Kea 13,796 ft. (4,205 m.)
160° 155°
20°

RUSSIA
Arctic Circle
Bering Strait
SEWARD PEN.
BROOKS RANGE
Pt. Barrow
ALASKA
Yukon River
Mt. McKinley 20,320 ft. (6,194 m.)
Tanana River
ALASKA RANGE
Iliamna Lake
BRISTOL BAY
ALASKA PENINSULA
Shelikof Str.
GULF OF ALASKA
Juneau
CANADA
BERING SEA
ALEUTIAN ISLANDS
0   250   500 Miles
0   250   500 Kilometers

GULF OF CALIFORNIA

114

UNITED STATES

- ⊛ National capital
- ★ State capital
- ● Major city
- ▬▬ International boundary
- ── State boundary

| 0 | 150 | 300 Miles |
| 0 | 150 | 300 Kilometers |

Projection: Albers Equal Area

CANADA

Lake of the Woods

Red Lake

Lake Superior

MICHIGAN

MAINE

Moosehead Lake

Mt. Washington 6,288 ft. (1,905 m.)

★ Augusta

Montpelier ★

N.H.

VT.

Concord ★

MINNESOTA

WISCONSIN

Lake Huron

Lake Ontario

ADIRONDACK MTHS.

Hudson R.

Lake Champlain

●apolis ● St. Paul

Mississippi River

Lake Michigan

Grand Rapids

★ Lansing

Detroit

Rochester

Niagara Falls

Buffalo

Syracuse

Albany

★

NEW YORK

Hartford ★

New Haven

Boston ★

MASS.

Cape Cod

Providence ●

R.I.

CONN.

Milwaukee

Madison ●

Susquehanna River

Newark

New York

IOWA

ILLINOIS

Chicago

Gary

Hammond

Toledo

Cleveland

Akron

Canton

Youngstown

PENNSYLVANIA

Pittsburgh

Harrisburg ★

Philadelphia

Trenton ★

N.J.

Camden

Dover ★

DEL.

● Des Moines

CENTRAL

LOWLAND

OHIO

★ Columbus

Dayton

★ Indianapolis

INDIANA

Cincinnati

Ohio River

Frankfort ★

WEST VIRGINIA

Charleston

MD.

Baltimore

Annapolis ⊛

Washington D.C.

DELAWARE BAY

Springfield ★

Wabash R.

Louisville

Richmond ●

Newport News

CHESAPEAKE BAY

ATLANTIC

OCEAN

Kansas City

St. Louis

East St. Louis

Jefferson City ★

Harry S. Truman Res.

KENTUCKY

Cumberland River

VIRGINIA

Norfolk

Roanoke River

Cape Hatteras

●as City

MISSOURI

OZARK PLATEAU

Knoxville

Nashville ★

Mt. Mitchell 6,684 ft. (2,037 m.)

Raleigh ★

Winston-Salem

APPALACHIAN MOUNTAINS

R.S. Kerr Res.

OZARK PLATEAU

ARKANSAS

Memphis

TENNESSEE

Tennessee R.

PLATEAU

NORTH CAROLINA

Lake Eufaula

Little Rock ★

Mississippi River

CUMBERLAND R.

Chattahoochee R.

Atlanta ●

Columbia ★

SOUTH CAROLINA

●as

Birmingham

GEORGIA

Rayburn ●ervoir

LOUISIANA

Toledo Bend Res.

ALABAMA

Jackson ★

Montgomery ★

Alabama R.

COASTAL PLAIN

MISSISSIPPI

★ Tallahassee

Jacksonville ●

Baton Rouge ★

Lake Pontchartrain

FLORIDA

Houston ●

New Orleans

Orlando ●

Cape Canaveral

GULF OF MEXICO

Tampa ●

St. Petersburg ●

Lake Okeechobee

N

Miami ●

Cape Sable

Key West

Straits of Florida

THE BAHAMAS

CUBA

95°

90°

85°

80°

75°

45°

65°

40°

35°

70°

30°

25°

The Northeast Region Today	
Agriculture	supplies local cities with produce such as vegetables, poultry, and dairy products; not a major producer of farm goods for the rest of the United States due to thin soils
Industry	manufactures electronic equipment, electrical machinery, tools, textiles, chemicals, and iron and steel; fishing, printing and publishing, chemical engineering, and medical research are additional industries
Movement	some Asian and Latin American immigrants arriving; people moving to the suburbs instead of living in the city; city governments trying to attract people and businesses back to urban areas by repairing old buildings and building new parks and malls

Source: *Geography and Development: A World Regional Approach*, James Fisher, 1989

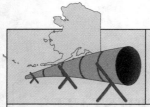

Alaska Native Claims Settlement Act, 1971

Native Alaskans (Inuits and Aleuts) lost much of their land as new people settled Alaska. Their ancestors' goals were to reclaim the lost land and govern themselves.

In the late 1960s oil was discovered in northern Alaska. The United States wanted to build a pipeline to carry the oil to the lower forty-eight states. Many native Alaskans had claimed the land on which the pipeline would run. The United States government passed the Alaska Native Claims Settlement Act in 1971. The native Alaskans received 44 million acres of land and $962.5 million. Twelve regions were set up in which the native Alaskans could govern themselves. These regions, or corporations, were organized around similar cultures. Each native was a stockholder of the corporation. In return, the United States government was able to build the oil pipeline.

Source: *Alaska: A to Z*, Angela Herb, 1993

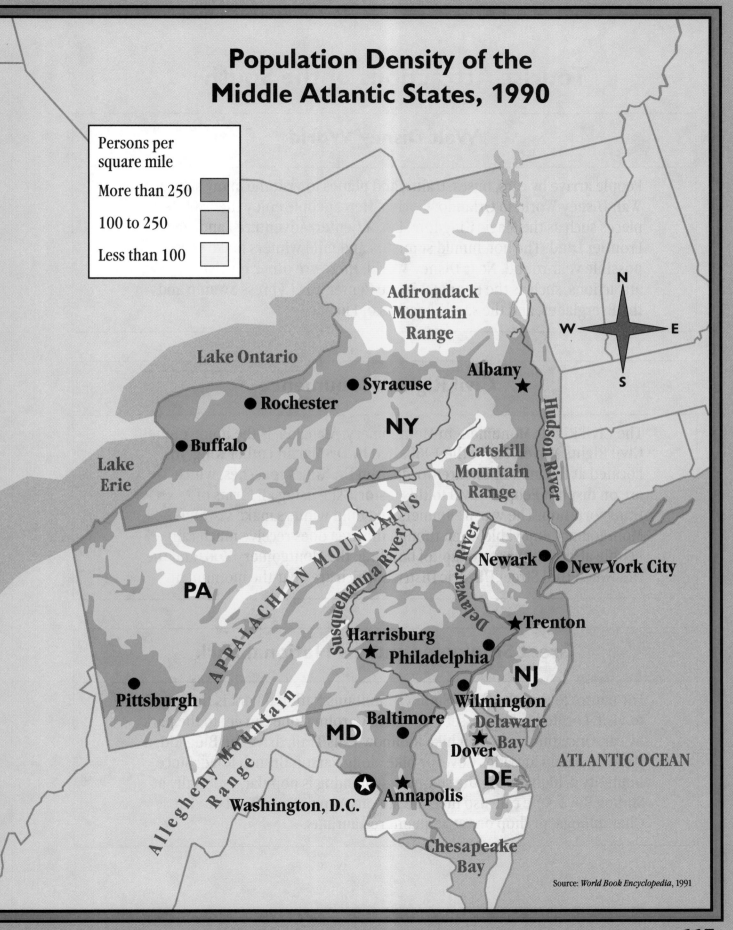

Population Density of the Middle Atlantic States, 1990

Persons per
square mile

More than 250

100 to 250

Less than 100

Adirondack
Mountain
Range

Lake Ontario

Syracuse

Albany

Rochester

NY

Buffalo

Lake
Erie

Catskill
Mountain
Range

Hudson River

APPALACHIAN MOUNTAINS

Susquehanna River

Delaware River

Newark

New York City

PA

Trenton

Harrisburg

Philadelphia

NJ

Pittsburgh

Wilmington

Baltimore

Delaware
Bay

MD

Dover

Allegheny Mountain Range

DE

ATLANTIC OCEAN

Washington, D.C.

Annapolis

Chesapeake
Bay

Source: *World Book Encyclopedia*, 1991

Tourist Attractions of the South

Walt Disney World

People arrive by cars, buses, trains, and planes to visit the many sites at Walt Disney World in Orlando, Florida. Here, people enjoy themselves at places such as the Magic Kingdom, Epcot Center, Adventure Land, and Frontier Land. The hot, humid summers and mild winters make visiting possible year-round. Near Disney World, there are other tourist attractions, such as the Kennedy Space Center, Big Cypress Swamp and the Everglades, and the sandy beaches of Florida.

Civil Rights Monument

The Civil Rights Monument in Montgomery, Alabama, is the home of the Civil Rights Movement of the 1960s, led by Dr. Martin Luther King, Jr. Located at the monument are words from Dr. King's speeches, which are on display. People travel to this historic site using all forms of transportation. The warm summers and mild winters make visiting Montgomery enjoyable, even though the area does receive much rain. People also visit the state capitol building, the Montgomery Zoo, and the Old North Hull Street Historic District, which are near the monument.

Great Smoky Mountains National Park

For nature lovers, the Great Smoky Mountains National Park is the place to visit. Located in Tennessee and North Carolina, one can get to the park by car. Once there, one can hike Rainbow Falls Trail, drive the Blue Ridge Parkway, camp at Cades Cove, or bike along Parson Branch Road. Since humidity is high during the summer, swimming is popular, especially at Sliding Rock. One can also hike the famous Appalachian Trail or visit Chattanooga, to shop or eat at quaint restaurants.

Kennedy Space Center

Kennedy Space Center is located in Cocoa Beach, Florida. It is the home of the rocket complex of the National Aeronautics and Space Administration (NASA). The climate is hot and humid during the summer and the winters are mild, with gentle coastal breezes yearlong. One can visit Spaceport, USA, or take a narrated bus tour. One can also see an IMAX film of space flight or visit the Brevard Museum of History and Natural Science. Cape Canaveral, located along the east-central coast of Florida, includes the Kennedy Space Center and also has tours. People arrive by planes, trains, buses, and cars to see this important center for science and research.

The Everglades National Park and Big Cypress National Preserve

The Everglades National Park and Big Cypress National Preserve are two of Florida's most beautiful physical features. These wetlands, located in southern Florida from Lake Okeechobee to Florida Bay and the Gulf of Mexico, thrive because of the hot summers and mild winters. People arrive by boat, car, and plane to see birds and animals in their natural environment. One can travel the Anhinga Trail, which is a boardwalk that leads into the swamp. It allows visitors to see junglelike plant life, crocodiles, alligators, manatees, huge turtles, and swamp birds. Visitors can also see Fort Jefferson National Monument, Biscayne Bay National Park, and the Florida Keys.

The Appalachian Trail

The Appalachian Trail extends from Mt. Katahdin in Maine to Springer Mountain in Georgia. Because the trail is so long, the climate may be different at each part of the trail that one is visiting. People can travel the trail by car. Along the Appalachian Trail, one can hike or ski. The area is also great for bird-watching.

Historical Peoples of the Southwest

Native American Group	Language and Customs	Food and Agriculture	Shelters and Clothing	Land Area and Climate
Navajo	• similar to the Apache language • sand painting for curing sickness	• hunting and gathering • herding of sheep, cattle, and goats • water source: rivers, lakes or streams	• hogans—houses made of earth and logs • garments made of animal skins	• northwestern New Mexico, Arizona, and southeastern Utah • warm, dry
Hopi	• Hopi language • dances in which snakes were released at the end to ask the Rain God to send rain	• sheepherding • farming: corn, squash, beans • water source: rivers, lakes or streams	• terraced, stone, and adobe apartments • cotton woven into a lightweight cloth	• northeastern Arizona • warm, dry
Pueblo	• Tewa, Zuni, and Keresan languages • dances to ensure good crops	• hunting • farming: corn and cotton • water source: river	• pueblo apartments—mud and stone square structures with many rooms • cotton woven into lightweight cloth	• northern Arizona, western New Mexico, along the Rio Grande • warm, dry

BosWash Megalopolis

Urban Area	Washington, D.C.
State Located	District of Columbia
Population of City	617,000
Major Industries	federal government, tourism, law and accounting, printing and publishing
Physical Features	located where the Chesapeake and Ohio Canal meet the Potomac River
Transportation Links to Other Cities	Dulles International Airport, National Airport, subways, superhighways

Urban Area	New York City
State Located	New York
Population of City	7,352,700
Major Industries	tourism, finance, apparel industry, printing and publishing
Physical Features	located on two islands and at the mouth of the Hudson River
Transportation Links to Other Cities	John F. Kennedy International Airport, Laguardia Airport, Penn Station, Grand Central Station

BosWash Megalopolis

Urban Area	Philadelphia
State Located	Pennsylvania
Population of City	1,647,000
Major Industries	finance, medicine, chemicals, clothing manufacturing
Physical Features	located along the Delaware River which flows into the Atlantic Ocean
Transportation Links to Other Cities	Philadelphia International Airport, Conrail trains, subways

Urban Area	Boston
State Located	Massachusetts
Population of City	577,830
Major Industries	tourism, clothing manufacturing, printing, defense, finance
Physical Features	located along the Charles River, Boston Harbor, and part of the Atlantic Ocean; the Blue Hills to the south and the Middlesex Falls to the north
Transportation Links to Other Cities	Logan International Airport in East Boston, subway system

BosWash Megalopolis

Urban Area	Baltimore
State Located	Maryland
Population of City	736,014
Major Industries	shipbuilding and ship repair, steel manufacturing, scientific research
Physical Features	located at the head of the Patapsco River near Chesapeake Bay
Transportation Links to Other Cities	Friendship International Airport, Harbor Tunnel Thruway, Francis Scott Key Bridge

California Water Projects

The Colorado River Aqueduct

The Colorado River Aqueduct serves the Los Angeles and San Diego area. Its source is the Colorado River at Hoover Dam.

The Los Angeles Aqueduct

The Los Angeles Aqueduct's source is the Owens River in the Sierra Nevada Range. This aqueduct supplies water to the city of Los Angeles.

The Hetch Hetchy Aqueduct

The Hetch Hetchy Aqueduct supplies water to the San Francisco Bay area. Its source is the Tuolumne River in the Sierra Nevada Range in eastern California.

Sources: *The World Almanac and Book of Facts*, Robert Famighetti, editor, 1995; *1996 Information Please Almanac*, Otto Johnson, editor, 1995; *Regional Geography of the United States*, Tom McKnight, 1992

Salt Lake City, Utah

History	
	Fleeing religious persecution in Illinois, the Mormons settled in Salt Lake City in July of 1847. The hot and dry desert climate in Salt Lake City made living there very hard. In the summer of 1848, crickets began eating the Mormons' crops, almost destroying them. The crops were saved by sea gulls from the great Salt Lake that swooped down and ate the crickets. In Salt Lake City, finding fresh water was a problem, also. Water was eventually piped in from other places for the crops and the people.
Today	
	Today, Salt Lake City is Utah's largest city. Industries such as steel and oil refining are important. Other economic activities that are important to Salt Lake City are copper mining, sugar refining, and high-tech firms, including research for the defense industry. The city is the center of the Mormon Church and serves as the center of government for Utah. It is the planned next site of the Winter Olympics.

Las Vegas, Nevada

History	
	Las Vegas, founded in 1905, was originally a mining town. It became a thriving city with the help of the Union Pacific Railroad. The hot and dry climate made living in the area hard. Because the water supply in Las Vegas was very low, many dams were built to bring water into the city. This helped the west-central region of the state develop into an important agricultural area. The dams brought many people to the city to work and live. Many hotels were built to house them. In 1931, gambling was legalized, leading to the growth of casinos.
Today	
	Today, tourism is the major industry in Las Vegas due to the hundreds of casinos, restaurants, and hotels located in the city. This has made Las Vegas Nevada's largest city.

Source: *Cities of the United States, The West,* Gale Research, 1989

Dams of the Southwest

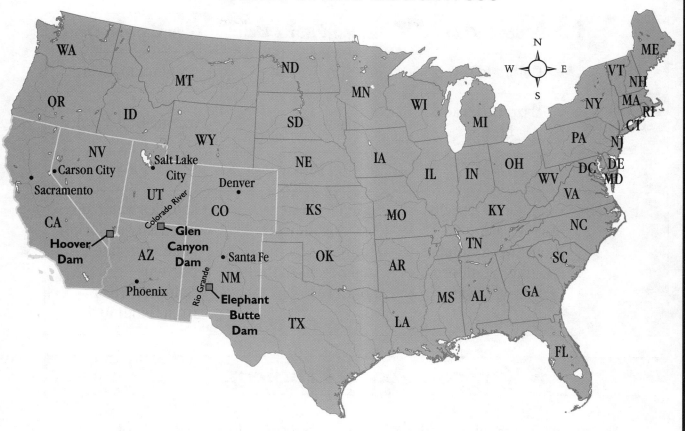

Hoover Dam

Facts about Dams of the Southwest

	Hoover Dam	Elephant Butte Dam	Glen Canyon Dam
River	Colorado River	Rio Grande	Colorado River
Location	Arizona-Nevada border	New Mexico	Arizona-Utah border
Year	1936	1916	1964
Purpose	• supplies electricity to the Pacific southwest; supplies southern California farms with water	• provides water for irrigation for cotton, grapes, and pecans in the southwest	• supplies water and electricity to Rocky Mountain states; prevents Colorado River from flooding

Environmental Problems

Several environmental problems have occurred since the construction of many dams in the United States. Controlling the flow of water is a problem. When a river's waters do not flow quickly, sand and soil settle to the bottom of the river. Fish that need a clean river bed to lay their eggs have no place to reproduce.

Another problem is the loss of important wildlife habitats. As water flows into the reservoir, the sand, stones, and dirt that it carried settle to the bottom. None of this material is emptied at the river's mouth. Sandbars and islands do not form in the river so many birds and fish have no place to feed or nest.

Finally, changes in plant life have affected the wildlife in the area. The changes in water level have destroyed natural growth patterns and types of plants. Many animals and fish that have depended on these plants as food are dying because their food sources are gone.

Sources: "The Trouble with Dams," Robert Devine, *Atlantic Monthly*, August, 1995
World Book Encyclopedia, 1991

Hawaii's Ecosystem

	Native to Hawaii	From Other Regions
Birds	• nene, or Hawaiian Goose (was near extinction but has been preserved) • Hawaiian stilt (endangered)	• crested honeycreeper (endangered) • Hawaiian coot (endangered) • myna, sparrow, cardinal
Animals	• hoary bat • Hawaiian monk seal	• dog, cow, goat, sheep, deer, horse, mongoose
Plants/Flowers	• coconut, breadfruit tree, bamboo, banana, sugarcane, sandalwood	• mango, papaya, avocado, pineapple, evergreen tree, eucalyptus

Hawaii's Main Products

Product	Location	Climate	Effect on the Land
Pineapples	• central Molokai • Lanai • Oahu (central valley)	• warm, moist climate and well-drained soil needed	• Streams that were shifted to support crops have prevented fish such as gobies from swimming to the ocean to reproduce.
Sugarcane	• Hawaii (northeast/southeast) • Maui (central/northwest) • Oahu (central valley)	• much rainfall needed (80–120") • temperature range between 75° and 86°	• Wildlife habitat was destroyed in order to build large plantations.

Sources: *World Book Encyclopedia*, 1991
Nature in Danger: Threatened Habitats and Species, Noel Simon, 1995
Hawaii Handbook, J.D. Bisignani, 1990

States of the New England Region

Maine

Vegetation	• pine, spruce, fir, sugar maple
Wildlife	• moose, whales, lobsters, black bears
Natural Resources	• soil: sand in coastal areas, clay in lowlands, and gravel-like soils in higher elevations; products include apples, potatoes, blueberries, and livestock • minerals: granite and limestone in southern Maine; slate in central Maine; other minerals such as copper, zinc, and silver • 90% of state is forested

New Hampshire

Vegetation	• ash, birch, cedar
Wildlife	• deer, black bears, ruffed grouse, beavers
Natural Resources	• soil: richest in valleys; products include dairy products, hay, apples, berries, maple syrup • minerals: gravel and sand used for roadbuilding materials • 80% of state is forested

Massachusetts

Vegetation	• eastern white and red pines, eastern Hemlock, pitch pine
Wildlife	• muskrats, deer, gulls, copperhead snakes
Natural Resources	• soil: richest in river valleys in peat; products include cranberries, eggs, livestock, dairy products • minerals: sand and gravel in east-central region; some granite • 60% of state is forested

Vermont

Vegetation	• ash, basswood, beech, birch
Wildlife	• white-tailed deer, black bears, beavers
Natural Resources	• soil: richest in river valleys; products include maple syrup, dairy products, hay, timber, livestock • minerals: granite, marble, and talc in Green Mountains; marble and slate in Taconic Mountains • 75% of state is forested

Connecticut

Vegetation	• hemlock, mountain laurel, bayberry
Wildlife	• white-tailed deer, foxes, minks, trout
Natural Resources	• soil: at low elevations, soil is dry and crops such as tobacco are grown; land unsuitable for agriculture; products include apples, poultry, eggs • minerals: sand, gravel, and traprock; state does not depend on its own resources for manufacturing because it lacks large mineral deposits • 60% of state is forested

Rhode Island

Vegetation	• ash, birch, dogwood, fresh/seawater seaweed
Wildlife	• otters, screech owls, ruffed grouse
Natural Resources	• soil: richest along Narragansett Bay; products include potatoes, apples, milk, greenhouse/nursery products • minerals: limestone, sandstone, and westerly granite in the southwest used for building materials • 60% of state is forested

Sources: *World Book Encyclopedia,* 1991;
Quick Facts About the USA, Nancy Hartley, 1994;
1996 Information Please Almanac, Otto Johnson, editor, 1995

Geographic Regions of California

The Mountains

California has two major mountain ranges. The Sierra Nevada Range lines the state's eastern border. The Coastal Range is located along the western edge of the state. Logging and mining are the major industries. There are no large cities in these areas, but resort towns around Lake Tahoe in the Sierra Nevada and Yosemite National Park draw many tourists. Because of its elevation, snow and rainfall provide the area with water.

The Central Valley

The California Central Valley is located between the Sierra Nevada and the Coastal Range mountains. Cities such as Sacramento, Fresno, and Stockton have grown because of this rich agricultural area. The Sacramento River and the San Joaquin River provide water for crops.

The Deserts

The Mojave Desert and Death Valley are located in southeastern and southern California. Cities are scarce, but Palm Springs, situated near the Mojave Desert, is a major resort town. Some mining also takes place in this area. Water is provided by the Colorado River and Owens Lake, in the Sierra Nevada.

The Coast

The California Coast is the home to cities such as San Francisco, Los Angeles, and San Diego. These are located on the land between the Pacific Ocean and the Coastal Range. Industries of this area include fishing, oil/gas production, computers, and entertainment. Water is drawn from aqueducts stemming from the Sierra Nevada Range and Hoover Dam along the Arizona-Nevada border.

Source: *America the Beautiful: California,* R. Conrad Stein, 1988

The Midwest Flood of 1993

States Affected	North Dakota, South Dakota, Minnesota, Wisconsin, Iowa, Nebraska, Kansas, Illinois, Missouri
Major Rivers Involved	Iowa, Mississippi, Des Moines, Racoon, Missouri, Minnesota
Crops Destroyed	soybeans, corn, wheat, sorghum
Damages	• 17,000 square miles of land flooded • 48 people killed • 26,000 people evacuated • 59% less corn and soybeans harvested in the United States • 21,219 homes damaged
Relief	Red Cross provided: • shelters in schools and firehouses offering places to sleep • bottled water for drinking • portable toilets (without water service flushing is impossible) • flood relief centers providing prepared meals for those with no homes during the crisis which lasted 16 weeks; Approximately 2.5 million meals were served. • over $30 million in relief effort spending

Sources: *The Mississippi Flood*, Karin Luisa Badt, 1993

Where Crops are Grown in the United States

Crop	Where It Is Grown
Corn	Illinois, Indiana, Iowa, Michigan, Minnesota, Missouri, Nebraska, Ohio, South Dakota, Wisconsin
Soybeans	Illinois, Indiana, Iowa, Kansas, Minnesota, Missouri, Nebraska, Ohio
Wheat	Colorado, Kansas, Missouri, Montana, Nebraska, North Dakota, Oklahoma, South Dakota, Texas

Corn

Corn is used by	Uses
People	• sweet corn, popcorn, cereals, salad dressing, margarine, corn-starch, baby food, and corn syrup • corn meal (made from ground corn) found in cornbread, tamales, and tortillas
Livestock	• corn feed which is made of ground shelled corn, ground ears of corn, and chopped corn plants after the ear has been removed • animal feed
Industries	• ceramics, explosives, construction materials, metal molds, paints, paper goods, textiles • medicines such as penicillin, vitamins, and antibiotics • fuel

Soybeans

Soybeans are used by	Uses
People	• cooking oils, margarine, mayonnaise, salad dressing, ice cream, cosmetics, baby food, cereals, pet foods, insect sprays, spray paint, tofu, soy sauce, candles, soaps
Livestock FEED	• animal feed
Industries	• adhesive tape, explosives, chemicals, textiles, fertilizer, disinfectants, fire extinguisher fluid, to provide firmness and protein in processed foods such as meats

Wheat

Wheat is used by	Uses
People	• flour, pasta, cereals, and bakery products
Livestock FEED	• livestock feed made from wheat germ and bran which remains after the white flour is milled • animal bedding which comes from the wheat stalk (straw)
Industries	• fertilizers, synthetic rubber, glue, fuel • baskets and hats from woven straw

National Geographic Historical Atlas of the United States, 1993; *World Book Encyclopedia,* 1991

GLOSSARY

A

arid: dry

author: a person who gathers the data for a map

B

bay: part of a sea that is partly surrounded by land, smaller than a gulf

boroughs: small parts of the state of Alaska; many boroughs make up the state of Alaska

C

canal: a waterway built to carry water from one place to another

canyon: a deep valley with steep sides, usually with a stream running through it

census: a counting, which happens every ten years, of the people living in the United States

clear-cut: cutting that uses a giant sawing machine to strip entire hillsides of trees

coast: land along the sea

coastal plain: the flat stretch of land along a shore

continental divide: an imaginary line running along the Rocky Mountains that divides the rivers that flow east from the ones that flow west

conurbation: a system of interconnected cities and other areas, both urban and suburban

counties: small parts of a state; many counties make up one state

D

dam: a piece of land that holds back the flow of a body of water

date: tells when data for map was gathered

diversity: having many differences

earthquake: a sudden shaking or shock inside Earth that causes movement on its surface

ecosystem: a group of living organisms that depend on one another and the environment in which they live

elevation key: a part of a map that helps the reader understand the shape of the land

elevation/relief: the height of the land above sea level

erosion: the gradual wearing away of the earth

evaporate: to change to water vapor

fault: a fracture or break in Earth's crust

fragile: easily destroyed or damaged

glacier: a huge mass of ice that slowly slides down a mountain

globe: a model that represents Earth as it looks from space

gorge: a deep, narrow opening between steep and rocky sides of walls or mountains

grid: a system of longitude and latitude; helps find places on the map

harbor: a deep body of water where ships can anchor

human region: an area that might be defined by the characteristics of its people, such as their language or government

index: an alphabetical list of the places shown on the map

Intermontane region: the area between the Rocky Mountains and the Pacific Coastal Range

irrigation systems: systems of channels, streams, or pipes used to bring water to crops

legend: tells the meanings of the symbols on a map

mouth: the place where a river empties into a larger body of water

old-growth: vegetation that has existed for a long time

orientation: tells what direction things are from each other on the map

orographic precipitation: rain that occurs on only one side of a mountain or hill due to warm air being forced upward by the rise in land

physical region: an area of the natural landscape that might be defined by its landforms, plant and animal life, natural resources, or climate

political map: a map that shows boundary lines

population density: the amount of people per unit of land area

port: a place with a habor where ships can anchor

region: areas of land that share features that make them different from other areas; features may be physical or human

rural: in, of, or like the country

scale: shows the measurement of distances and areas

seedcut: cutting that leaves a few trees in an area to provide seeds for a new crop

selective cut: cutting that clears mature trees from an area to make room for new trees to grow

semi-desert: a dry area with very sparse vegetation, often located between a desert and a grassland

shelterwood cut: cutting that leaves a few trees in an area for other trees that need shade to grow

source: tells who provided the data for the map

sparse population: a small number of people spread out over a large area of land

title: tells what the map is supposed to show; the main idea of the map

urban: in, of, or like a city

valley: the lower land that lies between hills and mountains

volcano: an opening in Earth's crust from which ashes and hot gases flow

waterfall: a stream that flows over the edge of a cliff

watershed: all the water that drains from a land area into a creek, lake, or river

wetlands: areas that have wet soils, such as swamps and marshes

INDEX

A. Read the definition of each word and look at its picture on the map. Notice that some words, such as *valley*, relate to land features. Some words, such as *bay*, relate to water.

1. Complete the chart below by writing the terms from the map in the appropriate column. An example has been done for you.

Land	Water
valley	bay
canyon	canal
coast	dam
glacier	harbor
port	mouth
volcano	waterfall

2. Under which of the above columns would you list continents? __Land__

3. Under which of the above columns would you list oceans? __Water__

B. Some of the different forms of land and water on Earth's surface are physical, or natural, features. Others are human-made features. Using the map and definitions shown, answer the following questions.

1. Which of the landforms and water forms are physical, or natural, features?

bay, canyon, coast, glacier, mouth, valley, volcano, waterfall

2. Which ones are human-made features?

canal, dam, harbor, port

3. Are there any that can be both formed by nature and made by humans?

dam

Lesson 1

ACTIVITY Identify land, water, natural, and human-made features.

Name That Feature

The BIG Geographic Question How can we learn to identify specific landforms and water forms?

In the article you read about the different ways Earth's land and water features can be represented. In the map skills lesson you learned the names of some specific land and water forms. Now create and play a game called "Name That Feature."

A. List the seven continents.

1. Africa
2. Antarctica
3. Asia
4. Australia
5. Europe
6. North America
7. South America

B. List the four oceans.

1. Arctic Ocean
2. Atlantic Ocean
3. Indian Ocean
4. Pacific Ocean

C. Unscramble each set of letters below to write the names of landforms and water forms.

1. coolvan — volcano
2. trop — port
3. mad — dam
4. aby — bay
5. laacn — canal
6. coats — coast
7. houtm — mouth
8. tellarfaw — waterfall
9. clearig — glacier
10. lavley — valley
11. cannyo — canyon
12. borrah — harbor

D. Using index cards and all of the words you listed in steps A, B, and C, make cards for a "Name That Feature" game. (Make sure students follow directions below.)

1. Make cards for each of the land and water features. On one side of the card, write the name and definition of the feature. On the reverse side of the card, draw a picture of the feature.

2. Make cards for the seven continents by drawing their shapes on one side of the card and writing their names on the reverse side.

3. Make cards for the four oceans by writing a description of the continents they are located near on one side of the card and writing their names on the reverse side.

E. Play the game by turning the picture-side of the cards face up and trying to name the feature. Turn the card over to check your accuracy. Challenge yourself by sorting all of the cards into land or water groups. (Make sure students correctly identify continents, oceans, landforms, and water forms and correctly sort cards into groups.)

B. Think about where you live in the United States. (Make sure students' answers accurately reflect the areas in which they live.)

1. Write the names of the states near your state that you think have similar features—such as beaches, mountains, forests, or weather—to your state.

_____   _____

2. If you could give your state, together with the ones you listed above, a "region" name, what would you call the region? Think about features the states share and write a region name below.

C. Looking at the map, think about what you know about the different areas of our country. What makes each area special or unique? How would you divide the United States into regions?

1. Write the names of the regions you would include. (Possible answers include the following.)

   a. Southwest   d. Northeast
   b. Midwest   e. Mid-Atlantic
   c. Northwest   f. Southeast

2. Make notes about where you will draw the borders for each region.

(Students' answers might reflect boundaries that indicate a similarity in the land, water, resources, climate, and/or vegetation of states in an area.)

D. Draw and label your regions on the map on page 10. (Make sure students draw their regions on the map.)

E. Compare your map with the regional maps in the article. (Answers will depend on the number of regions on the maps compared.)

1. Does your map have the same number of regions?

2. Which, if any, borders are alike, and which are different? Why do you think the same or different borders were chosen?

(Students' answers should describe differences in borders of specific regions.)

11

Lesson 2

ACTIVITY Identify features of regions.

What Makes Up a Region?

The BIG Geographic Question What are the characteristics of different physical and human regions of the United States?

From the article you learned that there are different ways to divide the United States into regions. The map skills lesson allowed you to create your own regions. Now identify the characteristics of specific human and physical regions of the United States and explain the similarities that make them a region.

A. Use the information in the article and Almanac to help you brainstorm a list of different kinds of regions.

1. Write the names of as many physical regions as you can find. For example, a physical region could be based on temperature.

(Possible answers include: mountain region, rainfall region, forest region, desert region.)

2. Write the names of as many human regions as you can find. For example, a human region could be based on farming.

(Possible answers include: Spanish-speaking region, technology region, entertainment region, tourism region.)

B. Choose any one of the physical or human regions from the article, map skills lesson, or this activity to read more about. Then complete the chart below. Write the name of your region in the space at the top. Show your region's location by outlining it on the map. List some features you think are characteristic of your region. (Students' answers should include the name of a region, its location, and characteristics.)

Region

Southeast (Human Region)

Location

Characteristics

- attractive to tourists (water sports)
- textile industry
- fun places to visit (Disney World)
- historical attractions

C. Get together with a partner and describe your region aloud. Include characteristics of your region and similarities to other regions. Challenge your partner to identify it and tell whether the region is physical or human. Then compare your maps. How are they similar? How are they different? Do your regions overlap?

(Students should be able to compare and contrast and identify common characteristics of their regions.)

13

A. Look at the map of United States industry in 1860. Answer the following questions.

1. What were the states involved in industry?

Maine, New Hampshire, Vermont, Massachusetts, New York, Pennsylvania,

Connecticut, New Jersey, Maryland, Ohio, Kentucky, Indiana, Illinois, Missouri,

and California

2. What were the main industries?

ironwork/steelwork, lumber, and textiles

B. Compare the 1860 map with the 1990 map. Answer the following questions.

1. Which states became industry centers in 1990 that were not in 1860?

Michigan, Tennessee, North Carolina, South Carolina, Alabama, Georgia, Florida,

Louisiana, Texas, Colorado, Arizona, Oregon, and Washington

2. What types of industries have developed since 1860?

printing/publishing, petroleum, electronics/computers, and chemicals

C. What do the maps show you about how regions of industry changed from 1860 to 1990?

Industry moved farther south and west in the United States, and

there were more and different types of goods being made in 1990.

Lesson 3

ACTIVITY
Find out the role geography plays in defining regions of the United States.

What A Place Makes

The BIG Geographic Question
In what way does geography affect the types of industries that exist in regions of the United States?

From the article you learned how two regions of the United States changed over time. The map skills lesson showed you how to use two maps to evaluate changes in the industry of various regions. Now find out how a region's location affects the type of industry that exists there.

A. Look at the maps on page 16 again. List the different types of industry that were in each region in 1990.

1. Northeast and Great Lakes Region

a. chemicals

b. electronics

c. printing and publishing

d. textiles

2. Mid-Atlantic and Southeast Region

a. chemicals

b. electronics

c. lumber

d. petroleum

e. textiles

---

3. West Coast Region

a. electronics/computers

b. lumber

c. petroleum

B. Look at your lists above and answer the following questions.

1. Near what important physical feature are most of the states with industry located? water (oceans and lakes)

2. Is the physical feature you listed above an important resource needed to help make many of the goods you listed in Part A?

Why or why not? No, because the products are made by machines and are not dependent on water power; Yes, because water is still a way that many of the products can be shipped to other countries.

3. Write three of the listed industries that are most important to you.
(Students' answers will vary, however electronics might be most frequently indicated since it is something students encounter daily.)

C. Select one of the industries listed above or one in the area where you live. Research how the industry you selected might be related to geography. Following are some questions to keep in mind while doing your research.

1. Where is the industry located?

2. What are the land, water, and resources (including people) in that location? Are they needed to make the product or good that this industry produces?

3. Was this industry important in the past? Is it important today? Why?

D. Make a one-page fact sheet about your selected industry to share with the class.

(Students' fact sheets should provide answers to the above questions in an interesting manner.)

A. List the six states shown on the map that make up the New England region. Then label them on the map.

1.	Connecticut	4.	Massachusetts
2.	New Hampshire	5.	Rhode Island
3.	Maine	6.	Vermont

B. Use information from the article and Almanac to identify the following. After you identify these places, locate and label them on the map.
(Make sure students have correctly labeled the following locations on the map.)

1. A name for the region of forests before European exploration

Eastern Woodlands (New England)

Trace this region on the map.

2. The mountains of eastern North America that run through

New England Appalachian Mountains

Draw these mountains on the map.

3. The ocean that the explorers crossed to get to North America Atlantic Ocean

Label this ocean on the map.

4. The country that sent Giovanni da Verrazano and Samuel de Champlain

to explore coastal New England France

Draw a line from this country to the New England region on the globe.

5. The country that sent John Smith to explore coastal New England England

Draw a line from this country to the New England region on the globe.

6. The harbor that Champlain mapped and described Plymouth Harbor

Label this harbor on the map.

7. The harbor that John Smith mapped and described Boston Harbor

Label this harbor on the map.

Lesson 4

ACTIVITY
Find out how New Englanders have used their land over the years.

Using the Land

The BIG Geographic Question
How have different groups in history changed the land in New England?

From the article you learned about the people who settled the New England region. In the map skills lesson you identified the place where European explorers who came to the region landed. Now find out how people used New England's land long ago.

A. Using the chart below, show how each historical group used the land and what it looked like afterwards. Include in your answers what you know about these groups in terms of farming, manufacturing, and the timber industry.

Group of People	Time Period	Ways the Land Was Used	Effects on the Land
Native Americans	500 A.D. to early 1600s	farmed crops such as corn, blueberries, strawberries, beans, squash	small settlement areas developed around fertile land
First European Settlers	1620 to early 1800s	farmed crops for food; timber used for ship-building and building homes; manufactured textiles	farms grew; more houses built; small towns developed
Immigrants of the Industrial Revolution	Mid-1800s	fields set up for mining iron ore; factories built for manufacturing steel; farmed to feed urban population	cities grew; farms still visible on landscape

---

B. Looking at the information collected on your chart, how has land use changed over time?

Land use changed from agricultural to industrial. However, some land is still being farmed to

support the urban population. Other land is being used in urban development.

C. What does the New England region offer, in terms of its own industries, that could benefit new businesses? Using the Almanac, list the industries on the chart below. Then write how you think each industry would benefit new businesses. (Possible answers include the following.)

Industry	Benefit
Agriculture	supports food supply for population
Iron/Steel	used to make many products; provides jobs
Medical research	brings in educated population; leads to cures for illnesses
Printing/Publishing	provides jobs for population
Electronic equipment	leads to development of new technology; provides entertainment
Fishing	food supply for population/export
Electrical machinery	allows for industries in the area to work more efficiently

D. Imagine that you are at a meeting of New England governors. You are to write a letter to businesspeople across the United States. You are trying to get them to do business in the area. What would you say to attract them? Write your letter on a separate piece of paper and share it with the class.
(Make sure students' letters include features—old and new—of New England that might attract businesses to come there.)

A. Match each of the following words with one of the clues given below and reveal an acronym for map elements. The circled letters form the acronym.

Legend   Grid   Author   Date   Source
Title   Index   Orientation   Scale

1. Tells what the map is showing (T) i t l e

2. Tells what direction things appear in relation to each other
(O) r i e n t a t i o n

3. Tells when the data was gathered (D) a t e

4. Tells who gathered the data (A) u t h o r

5. A box that shows the meanings of the symbols (L) e g e n (d)

6. Used to compare map distance to actual distance (S) c a l e

7. Lists alphabetically the places shown on the map (I) n d e x

8. Helps find the location of a particular place on a map (G) r i d

9. Shows who provided the data (S) o u r c e

10. What is the acronym? T O D A L S I G S

B. Look at the New England regional map. Identify or describe the map elements using the acronym TODALSIGS.

1. T New England Region, 1996

2. O The New England coast, north/south along Atlantic Ocean

3. D 1996

4. A SRA/McGraw-Hill

5. L Contains symbols for mountains, rivers, and state capitals

6. S Provides a way to measure distance between places

7. I Lists the locations of the New England states

8. G Shows a system of vertical and horizontal lines that help locate places

9. S John Edwards and Associates

29

Lesson 5
ACTIVITY   Make a map of the physical and human elements of the New England region.
A Map of New England

The BIG Geographic Question   What are some of the key physical and human elements of New England?

From the article you learned about physical and human features of New England. In the map skills lesson you learned an acronym to help you recall map elements. Now make a map showing physical and human elements of the New England region.

A. Look at the features listed on the chart below. Choose at least one of these New England resources to research in the Almanac. Take notes as you do your research. (Make sure students choose at least one of the features to research.)

Feature	What to Research	Almanac Notes
Mountains	name, where located	
Rivers	name, where located, mouth	
Vegetation	forest lands	
Natural Resources	types of natural resources	
Wildlife	types of wildlife	

B. Decide how you will show the information you have noted on a map. Use a sheet of scrap paper to plan and sketch your map. Then draw your map of the New England region in the space below. Give your map a title.

(Students' maps should reflect the information they gathered on the chart. They may choose to use pictures, labels, graphs, or shapes to show their information.)

31

A. Look at the elevation key to answer the following questions.

1. Write the elevation of the land shown in yellow.
   152 meters   500 feet

2. Which elevation is higher, the light green area or the orange area?
   the orange area

B. Look at the map of the Middle Atlantic region.

1. What is the elevation of most of the land along the Atlantic Ocean?
   0 meters, 0 feet, sea level

2. Describe how the land changes as you travel from west to east.
   (Students should describe the gradual change from mountains of around 1,000 feet in elevation to land at sea level at the Atlantic Ocean.)

C. Find Buffalo, New York, on Lake Erie. Draw a line from west to east going from Buffalo to Albany, New York.

1. What two colors do you pass through? What are the elevations?

   a. Elevation color   b. Elevation in feet
   orange   1,000 feet
   yellow   500 feet

2. What is the name of the mountain range north of the line that you drew?
   Adirondack Mountain Range

3. Continue your line from Albany, New York, by tracing the Hudson River south to New York City. What is the elevation in feet? 0 feet or sea level

4. Do you think the route from Buffalo to Albany was a good route for the Erie Canal? Why? (Students should mention that the route followed the least change in elevation, which made it an easier route to build the Erie Canal.)

35

Lesson 6
ACTIVITY   Find out how the geography of a region influences where people live.
Why Live There?

The BIG Geographic Question   How do the land and water of a region affect settlement patterns?

In the article you read about the diverse physical features of the Middle Atlantic region. In the map skills lesson you looked at the elevation of this region. Now find out more about how geography affects people's locations and occupations.

A. Look in the Almanac for the population density map for the five states of the Middle Atlantic region.

1. Where is the population the greatest? Why?
   along the eastern coast; water provides a way for traveling and transporting goods

2. Which two states have the greatest rural areas?
   Pennsylvania   New York

3. What ocean is to the east of this region? Atlantic Ocean

4. What two Great Lakes border the western side of this region?
   Lake Ontario   Lake Erie

5. List three major rivers in this region.
   Hudson River   Delaware River   Susquehanna River

6. List two big bays on the east coast.
   Chesapeake Bay   Delaware Bay

7. List mountains in this region.
   Appalachian Mountains; the Adirondack and Catskill Mountains are ranges in the Applachians

---

B. Think about the geography and the population of the Middle Atlantic region.

1. What do you notice about the relationship between the features of the land and where people live? The population density is the greatest along the eastern shore and the lowest in mountainous areas.

2. How do you think people decide where to live?
   (Students' answers should show an understanding that people have settled in areas with the best access to water for agricultural use, household use, and transportation and where mountains are not an obstacle to movement. Also, water is a source for recreational activities such as swimming, skiing, and fishing.)

C. Now, using clay or papier-mâché, make a model of what you have learned about the Middle Atlantic region. Be sure to include the major landforms and waterways and mark the cities.

Here are a few hints:

- Glue yarn or string to your model to mark the major waterways.

- Use extra clay or papier-mâché to build up the mountain areas.

- Use flags made of toothpicks and small pieces of paper to mark the mountain ranges. Write the elevations on your flags.

- Make your own map key and symbols to show information. See an example below.

Sample Map Key
★ state capital
○ ocean
▲ mountains

1. What is the name of the closest major city southwest of New York City? _____
   Philadelphia

   Which map showed you New York City with other major cities? *Map 2*

2. What building is located slightly southwest of the New York Public Library?
   Empire State Building

   Which map shows the building? *Map 1*

3. In what region is the state of New York located? *Middle Atlantic*

   Which map show you the state of New York in a region? *Maps 2 and 3*

4. What street runs along the west side of Central Park? *Broadway*

   Which map shows the street? *Map 1*

5. Look at the three maps and answer the following questions.
   a. What river borders the east side of Manhattan? *East River*

   b. What river borders the west side of Manhattan? *Hudson River*

   c. Which one of these rivers has two tunnels connecting New York and

   New Jersey? *Hudson River*

   d. Name three boroughs, or sections, within New York City that are
   shown on the map.
   Queens

   Manhattan

   Brooklyn

6. Put an *X* in the box next to the map that shows a larger part of the United States.

   ☐ Map 1   ☐ Map 2   ☒ Map 3

Lesson 7

ACTIVITY Compare and contrast the five major urban regions of the BosWash Megalopolis.

Megalopolis Mania

The BIG Geographic Question How are the urban regions of the BosWash Megalopolis alike and different?

From the article you learned what the BosWash Megalopolis is. The map skills lesson showed you New York City and the megalopolis on different scale maps. Now compare and contrast the five major urban areas that make up the BosWash Megalopolis.

A. On the chart below, write down some things you know about each urban area of the BosWash Megalopolis. Use the article and map skills lesson to help you with this information. One urban area has been done for you.
(Students' answers should accurately describe each megalopolis urban area.)

Megalopolis Urban Area	Where It's Located	Physical Features
Washington, D.C.	between Virginia and Maryland	• near Potomac River

---

B. Using the Almanac information, complete the following to learn how the megalopolis urban areas are alike and connected.

1. All five urban areas have populations

   _____ under 500,000   ___X___ over 500,000.

2. All five urban areas have _____international_____ airports that connect

   them with each other and cities around the _____world_____.

3. All five urban areas are located near waterways, such as the

   _____Atlantic_____ Ocean or a _____river_____.

C. Use the information from your chart and the Almanac to create a 2- or 3-dimensional cityscape of the BosWash Megalopolis. A cityscape is a representation of a city's landscape. Plan your cityscape on a scrap piece of paper. Include all five urban areas and show that they are connected to form one region. Here are some suggestions.
(Students' cityscapes and descriptions should reflect features of the BosWash Megalopolis urban areas.)

- Draw and color or paint different kinds of landscapes on posterboard to represent each urban area's scenery. The scenery might include highways, railways, airports, seaports, and so on.
- Use index cards to draw and cut out the faces of features such as buildings, trees, or airplanes.
- Prop up all of your features by taping one end of a piece of cardboard against their back sides and the other end to the landscape.
- Write a description of the BosWash Megalopolis. Include information about each urban area's physical features and the transportation routes that connect them.

_____

_____

_____

_____

_____

A. Look at the maps of southern Florida. Put an *X* on the following places on each map.

Map A
1. Lake Okeechobee
2. Big Cypress Swamp
3. Everglades
4. Kissimmee River

Map B
5. Florida Panther National Wildlife Refuge
6. Everglades National Park
7. Big Cypress National Preserve
8. Florida Trail

B. Use both maps to complete the following.

1. Indicate on which map you can find each of the following and whether each is natural or human-made.

   a. Big Cypress Swamp? *Map A; natural*

   b. Orlando? *Map B; human-made*

   c. Big Cypress National Preserve? *Map B; human-made*

   d. an interstate highway? *Map B; human-made*

   e. a canal? *Map B; human-made*

   f. rivers? *Map A; natural*

2. Look at Map B. Where are most of the cities located?

   Most cities are located along the coast and along the edges of the Big Cypress Swamp and the

   Everglades.

3. Draw the information from Map B onto Map A. (Make sure students complete the following.)
   a. Place the cities, roads, and canals on Map A.
   b. Add the boundaries to show Big Cypress National Preserve and Everglades National Park.
   c. Are any human-made features such as cities located inside Big Cypress

   National Preserve and Everglades National Park? *No*

4. Now look at Map A. Describe how much of the Everglades and Big Cypress Swamp has been preserved by humans and what has happened to the remainder of the wetlands.

   (Students' descriptions should indicate that approximately one third of each wetlands environment has

   been preserved and the remainder has been drained for farming and development.)

Lesson 8

ACTIVITY Find out the pros and cons of changing fragile wetlands.

Changing Environments

The BIG Geographic Question Should people develop or preserve fragile ecosystems like the wetlands in southern Florida?

From the article you learned that the wetlands of southern Florida are very fragile but important environments. In the map skills lesson you looked at some human-made versus natural features in southern Florida. Now find out what different groups want to do about the future of the wetlands.

A. Indicate what you know about the following wetland features. Add other features from the article or your own research. Write *N* for natural, *H* for human-made, *+* for positive impact on the wetlands, and *–* for negative impact on the wetlands. (Make sure any listed features are correctly identified as natural or human-made and positive or negative.)

Features	Type	Impact
parking lots	H	–
paved roads	H	–
cypress trees	N	+
sawgrass	N	+
canals	H	–
alligator holes	N	+

B. Imagine you have to present the viewpoint of a group that has a special interest in the wetlands of southern Florida. Choose one of the groups below. You might want to talk with an adult friend or family member to see what they know about the groups listed. Best of all, if you know someone who fits one of these descriptions, you might want to ask that person the questions in section C below. Circle the group whose ideas you want to represent.
(Make sure students understand the groups they circle.)

1. Land developers—want to build new condominium complexes or vacation areas
2. Environmentalists—want to preserve the environment
3. City dwellers moving out of the city—want to move to a quiet place to enjoy nature and get away from the crowded city
4. Farmers—want land to farm and water to irrigate crops
5. Senior citizens retiring to southern Florida—want to move to a place with warm weather and plenty of recreational activities close to home

C. See how many of the following questions you can answer about your group. Do research if you need more information.
(Students' answers should be consistent with the general interests of their selected groups.)

1. Does your interest group want to preserve or develop the wetlands?

   _____

2. What has happened to the wetlands in the past that you think is good?

   _____

3. What has happened to the wetlands in the past that you think is bad?

   _____

4. What are your group's plans for the wetlands?

   _____

5. Why would following your group's plans be a good idea?

   _____

D. Present to a friend or family member your group's ideas on the issue of preserving versus changing the wetlands. Use books, pictures, and maps to enhance your presentation.
(Students' presentations should reflect the interests of the groups they have chosen to represent.)

A. Look at the maps and answer the following questions.

1. In 1860, were most of the cities located in the northeastern states or the southeastern states? _northeastern states_

2. In 1990, where were most of the cities located? _in the northeastern and southeastern states_

3. What cities were important industrial centers in 1990 that were not important in 1860? _Nashville, Charlotte, Birmingham, Atlanta, Jacksonville, Orlando, Tampa, St. Petersburg, Miami, New Orleans, Houston, Dallas, Phoenix, San Diego, Los Angeles, Portland, and Seattle_

4. To what areas did industry spread in 1990? _the southeast, southwest, and northwest coast_

B. Looking at the maps and your answers to the above questions, draw some conclusions about how the location of major industrial centers has changed over time.

1. Read the statement: Manufacturing moved from the industrial northeast to the "New South." Explain why you agree or disagree with this statement.

(Students may agree with the statement because the map shows the growth of industrial centers in the southeast in 1990. They may disagree because the statement does not address growth of industrial centers in the west, which the map also shows.)

2. Why do you think most of the industrial centers in 1860 and 1990 were mainly on the east and west coasts?

The oceans and rivers provided water, which could be used for transportation and to power factories in 1860 and for shipping in 1990.

Lesson 9

ACTIVITY Find out what makes a good tourist attraction.

"Hot" Tourist Spots

The BIG Geographic Question   Why has tourism become such big business in the Southeast?

From the article you learned about the South and how it has changed. The map skills lesson showed you how the location of industrial centers changed over time. Now find out why tourism has become such a big business in the Southeast.

A. Circle one of the following tourist attractions in the Southeast, or write the name of another southeast destination that you would like to know more about. (Make sure students know what each of the below attractions are.)

- (Civil Rights Monument)
- Walt Disney World
- Kennedy Space Center
- Smoky Mountain National Park
- The Appalachian Trail
- The Everglades National Park

B. Answer as many of the following questions as you can about the tourist attraction you selected. Use a pencil so you can change your answers later by checking them against the Almanac information. (Students' answers should be reasonable for the attraction selected. An example is provided.)

1. Where is your tourist destination located? _Montgomery, Alabama_

2. What is the climate like there? _warm summers, mild winters_

3. What are some specific things to do there? _read words from Dr. Martin Luther King, Jr.'s speech and learn about the Civil Rights Movement_

4. What else is nearby that is of interest? _the state capitol building, Old North Hull Street Historic District, the Montgomery Zoo_

5. How could you travel there? _car, train, bus, plane_

C. What do you think are important features of a tourist attraction or destination? Use the chart below to indicate your rating by marking a + if the feature is important. Mark a – if the feature is not important. Then ask two classmates or family members for their ratings. (Make sure students indicate which features are important to them.)

Feature of Tourist Attraction	Your Rating	Person 1 Rating	Person 2 Rating
places to camp nearby			
places to eat nearby			
cost to get in			
location			
weather and climate			
other tourist sites nearby			
educational			
fun			
variety of activities			
shopping			

D. Design a travel brochure for that tourist attraction. Include pictures, a map, or any other graphic that will add interest to your brochure. The brochure should include reasons why people should visit the site. Use ideas from the chart above. Then look over your travel brochure and those of your classmates. What common feature is important in attracting people to the Southeast?

(Students' brochures and answers should reflect that the Southeast has many fun, historical, and natural features that tourists find attractive, but people are probably most attracted by its sunny weather, warm climate, and many beaches.)

A. Look at the map and circle the following. (Make sure students accurately circle the following.)

1. Illinois   3. Nebraska   5. Missouri
2. Iowa   4. Arizona   6. Map key

B. Use the map and map key to answer the following questions.

1. a. What does a red dot mean? _10,000 acres of corn_

   b. What does a green dot mean? _20,000 acres of wheat_

2. a. Name at least five states where corn is grown. _Illinois, Iowa, Wisconsin, Ohio, Nebraska, Kansas, Missouri, Minnesota, and South Dakota_

   b. How do you know corn is grown in the states you named? _The red dots and Corn Belt label indicate where corn is grown._

C. Answer the following.

1. How many days do you see in Arizona? _3_

2. What color is each dot? _There is a red dot, a light green dot, and a dark green dot._

3. How many acres of corn are grown in Arizona? _10,000 acres_

   How do you know? _The map key shows that one red dot is equal to 10,000 acres of corn._

4. Is more corn grown in Arizona or Illinois? _Illinois_

   Explain how you know. _There are many more red dots in Illinois than in Arizona._

D. Write a few sentences comparing the amount of corn grown in Ohio to the amount of corn grown in Missouri.

Ohio and Missouri seem to grow almost the same amount of corn. Ohio's corn seems to be grown all over the state, whereas Missouri's corn seems to be grown mostly in the northern part of the state.

Lesson 10

ACTIVITY Find out how a natural disaster affects people—directly and indirectly.

Floods, Farmers, and Food

The BIG Geographic Question   How does a natural disaster in one region affect people in another region?

From the article you learned that the Midwest is known as the industrial and farming heartland of the United States. A dot map in the map skills lesson helped you locate corn and wheat belts in the Midwest. Now find out how something that happens in one region of the United States can touch the lives of people in another region and sometimes all over the world.

A. List some natural disasters that you think could hurt a farmer's crops.

1. _drought_   4. _insects_
2. _floods_   5. _fires_
3. _tornadoes_   6. _hurricanes_

B. The Great Midwestern Flood of 1993 was a natural disaster. Try to answer as many of the following questions as you can about the flood. Use a pencil so you can change your answers later. Look in the Almanac to check and adjust your answers.

1. List the nine states that were directly affected by the flood. _North Dakota, South Dakota, Nebraska, Kansas, Minnesota, Wisconsin, Iowa, Illinois, Missouri_

2. Which two rivers were the main source of the flood waters? _the Mississippi River and the Missouri River_

3. What were some of the crops grown in the flood area? _corn, wheat, and soybeans_

C. Many people felt the effects of the Great Flood of 1993. Read each effect. Mark D if it directly affected the people; mark I if it indirectly affected the people. (Students' answers might look like the following.)

Effect	People in Flood Area	People Outside Flood Area
1. Flood waters and river bottom sand covered farm fields.	D	I
2. Homes and other structures were filled with water and mud.	D	I
3. Roads closed and bridges washed away.	D	I
4. Drinking water was contaminated.	D	I
5. Farmers couldn't harvest crops or plant new crops.	D	D
6. Volunteers helped flood victims.	D	D
7. Certain foods were in short supply in the grocery stores.	D	D

D. Use the information from the article, the map skills lesson, the Almanac, and the information you collected above to create a news broadcast about the flood. Describe the areas directly affected by the Great Midwestern Flood of 1993. Describe the physical features involved and the problems caused by the flood. Write your broadcast below. (Students' broadcasts might include the following.)

- the 9 states affected by the flood
- the major rivers that flow through the Midwest
- the crops grown in the area
- the damage done to homes and farm foods
- major roads and bridges affected by the flood waters
- how flooding affects people inside and outside the flood area

A. Look at the map and complete the following. (Make sure students underline or circle the correct
   1. Underline the names of the states in the northern spring wheat states for each question below.)
      belt.
   2. Circle the names of the states in the southern winter wheat belt.

B. Answer the following questions about where the different kinds of wheat are grown.

   1. Which two states grow the most wheat? _Kansas and North Dakota_

   2. Which states produce both spring and winter wheat? _Washington, Oregon, Idaho, Montana,_
      and South Dakota

   3. Which states are minor producers of winter wheat? _Wisconsin, Iowa, Indiana, Michigan, Ohio,_
      Tennessee, North Carolina, South Carolina, Georgia, New York, Pennsylvania, Virginia, Kentucky,
      California, Arizona, Utah, New Mexico

   4. Is there more winter wheat or spring wheat grown in the United States? _winter wheat_

   5. Which three spring wheat producing states grow the smallest amount of spring wheat?
      Washington, Oregon, and Idaho

   6. Where in North America is the most spring wheat grown?
      northern United States and southern Canada

Lesson 11

ACTIVITY   Find out about crops that are grown on
   the plains and how they are used.

Growing Grains

The BIG Geographic Question   What agricultural products do we get from the Midwestern plains?

In the article you read about how people started growing wheat on the plains and what effect it had on the land. The map skills lesson showed where wheat is grown in the Midwest today. Now find out more about crops grown in the Midwest and the uses people have for them.

A. Use the Almanac to find where wheat, corn, and soybeans are grown in the Midwest. List the crops and the Midwestern plains states where each crop is grown.

Crop	Where Grown
Wheat	Missouri, North Dakota, South Dakota, Montana, Kansas, Oklahoma, Texas, Nebraska, and Colorado
Corn	Iowa, Illinois, Nebraska, Minnesota, Indiana, Ohio, Wisconsin, Missouri, Michigan, and South Dakota
Soybeans	Iowa, Illinois, Minnesota, Indiana, Missouri, Ohio, Nebraska, and Kansas

B. Find out how each crop is used. First look through your home for products made from each crop. Then find out about other uses. Try to find at least four uses for each crop and list them below. At least two uses should be something other than food items.
(Possible answers include the following.)
Wheat: bread, pasta, cereal, livestock feed, straw for baskets or hats, fertilizer, plywood glue, fuel
Corn: animal feed, sweet corn (human food), cereal, tamales, tortillas, popcorn, margarine, corn syrup, cornstarch, fuel (gasohol and ethanol), medicines, vitamins, paint, paper goods, ceramics, construction materials, textiles
Soybeans: animal feed, baby foods, cereals, tofu, processed meats, soy sauce, cooking oil, mayonnaise, salad dressings, candy, ice cream, fertilizer, fire extinguisher fluid, insect spray, paint, candles, disinfectants, soap, tape, drugs, explosives, cosmetics, textiles

---

C. On a separate sheet of paper, make a chart like the one below. List the crops you learned about. List the Midwestern states where they are grown, and picture or list some uses of each crop.   (Check students' charts to make sure they have supplied the correct place where grown and uses for each crop listed.)

Crop	Where Grown	Uses
Wheat	North Dakota	bread

D. Choose one Midwestern plains crop. Write about how important this crop is in people's everyday lives. What might happen if flooding or storms destroyed much of this crop during a growing season?

(Students' answers should show an understanding of the crop's importance and the effect its destruction would have on people.)

A. Look at the migration map. Trace these three states on the map. (Make sure students have correctly outlined the three states.)
   1. California (CA)
   2. Texas (TX)
   3. Florida (FL)

B. Use the map to complete the following.
   1. What is the source of this migration map?
      United States Department of Agriculture

   2. Read the map's title. What do the three states represent on the map?
      The three states represent the starting points of the farm workers' migration routes.

   3. In which direction did the workers migrate?
      They mostly migrated north.

   4. Why do you think the farm workers migrate to the north in the summer?
      As the seasons change, the workers follow the crops north to get another job.

   5. Suggest other reasons why people might move from one place to another.
      Ideas might include better weather, family connections, and seasonal job opportunities as pull factors.
      Push factors might include war in their home country, lack of jobs, and not enough food.

   6. Choose one migration route and circle it on your map. Describe how the route moves and changes. Be sure to use direction words (north, south, east, west) and the state names in your description.
      (Students should clearly mark a route on the map. Their written descriptions should be clear and
      correctly use cardinal and intermediate directions and state names.)

Lesson 12

ACTIVITY   Create a cultural collage of the Southwest.

A Southwestern Collage

The BIG Geographic Question   Who are the people of the Southwest, and what are their customs?

From the article you learned about different cultures that lived in the Southwest long ago. The map skills lesson showed you how to use a map to understand movement patterns of migrant workers. Now make a collage of pictures and words about the cultures of the Southwest, past and present.

A. Write the names of the states in the southwestern region of the United States.

   Arizona, New Mexico, and Texas

B. In the box below, draw a simple map of the Southwest. Label the states. (Make sure students' maps correctly show the three states of the Southwest.)

C. Use the article and the Almanac to find out more about the Southwest. As you learn about the region, take notes about the different people who live there and their cultures.
(Students' notes should reflect the various characteristics of the Southwest.)

Characteristics of	Notes
People	
Food	
Language	
Clothing	
Customs	
Climate	
Land	
Water	
Buildings	

D. Look through magazines and newspapers. Cut out pictures or words that represent southwestern cultures. You might want to draw pictures and write words of your own to add to your collection.
(Make sure the pictures and words students select are characteristic of the Southwest.)

E. Put your drawings, pictures, and words together to make a large collage about the Southwest. Share your collage with the class.
(Make sure students' collages are representative of the Southwest.)

A. Look at the climograph and complete the following. (Make sure students are following directions.)

1. Find and touch *January* at the bottom of the graph.

2. Move your finger up until you touch the colorful, horizontal line on the graph.

3. Move your finger to the left until you touch the column of numbers that stand for temperatures. The number nearest your finger is close to the average temperature in Phoenix in January. That temperature is about 52°F.

4. Down the right side of the graph, circle the number that stands for inches of rainfall.

B. The bars on the graph show inches of rainfall in Phoenix. To find (Make sure students are following directions.) **the average rainfall in Phoenix in January, follow these steps.**

1. Find and touch *January* at the bottom of the graph.

2. Move your finger up until you touch the top of the bar.

3. Move your finger to the right until you touch the column of inches of rainfall. The number nearest your finger is close to the average rainfall in Phoenix in January. That amount is almost 1 inch.

C. Answer the following questions using the climograph.

1. What is the average temperature in Phoenix in July? _91°_

2. What is the average rainfall in Phoenix in July? _almost 1 inch_

D. Compare Phoenix's average temperature and rainfall in January and July. Would you say that Phoenix has a hot and dry climate, a hot and wet climate, a cool and wet climate, or a cool and dry climate?

January _cool and dry_

July _hot and dry_

E. Examine the temperatures and inches of rainfall for Phoenix for all twelve months. Write a brief statement explaining what you see.

Overall, Phoenix appears to get very little rainfall and warm temperatures.

Lesson 13

ACTIVITY   Find out the positive and negative effects of building dams in the Southwest.

Damming Wild Rivers

The BIG Geographic Question   What are the positive and negative effects of building dams?

In the article you read about the Grand Canyon in the Southwest. In the map skills lesson you discovered how warm and dry states in the Southwest can be. Now study the efforts people have made to bring water to this area through the building of dams.

A. Dams are an important land feature found in many places. Put an X beside each statement below that describes how a dam might be used.

- _X_ To control the water flow of a river so that areas along the river get about the same amount of water
- _X_ To conserve water for dry periods that may occur throughout the year
- _X_ As a source of electrical power

B. Look at the map on page 125 of the Almanac and complete the chart below.

Dam	Location	River
Elephant Butte Dam	New Mexico	Rio Grande
Glen Canyon Dam	Arizona – Utah border	Colorado River
Hoover Dam	Arizona – Nevada border	Colorado River

C. Find out some reasons that these dams were built. List them on the left side of the chart below. Then find out some problems that have resulted from building these dams in a dry, semi-desert region. List them on the right side of the chart.

Building Dams

PROS (Good Effects)	CONS (Bad Effects)
control flooding	change the environment
produce electrical power for industry, farms, and homes	destroy natural plant life
produce irrigation water	possibly harm wildlife
produce drinking water	block natural flow of river
produce recreational areas	sand, stones, and dirt settle to the bottom of the reservoir and are not emptied at river's mouth
	no place for fish eggs to mature
	loss of wildlife habitats and plant life

D. Make a time line that shows the building of major dams in the Southwest. For each dam, mark the year it was completed and the river on which it was built.

```
        1916    1936          1964
   1905  1915  1925  1935  1945  1955  1965  1975
          |            |                  |
   Elephant Butte Dam  Hoover Dam    Glen Canyon Dam
   on Rio Grande       on Colorado River  on Colorado River
```

E. Think about what you have discovered about dams. Write about the environmental effects—both good and bad—of building dams in the Southwest. Do you think dam building does more good than harm, or more harm than good? Tell what you think and why.

(Students' paragraphs should include the pros and cons of building dams, their opinions on whether dam building does more harm than good, and facts to support their opinions.)

A. Study the 1860 map to find answers to these questions.

1. The routes of the Mormon Trail, the Oregon Trail, and the Sante Fe Trail each followed a river for part of the journey. Why do you think they did this?

The rivers were a reliable source of water transportation.

2. What major obstacle did each of the trails cross? _the Rocky Mountains_

3. The California Trail split from the Oregon Trail at Fort Hall. Which trail do you think was an easier route to follow?

Going to California meant crossing the mountains and the desert; going to Oregon meant crossing mountains only. Both routes were very difficult.

B. Study the historical 1860s and contemporary 1990s maps to find answers to these questions.

1. People travel across the country a little differently today than they did in 1860. What does the 1990s map tell you about the West today? _There are several_ roads that go through or around the mountains and desert. These roads make travel easier.

2. Why would some routes on the contemporary map be the same as the routes on the historical map? _Some of the historical routes have proven to be good routes_ to travel. Roads have been built along these routes.

3. Which of the communities on the historical map are also on the contemporary map? _Portland, Sacramento, Salt Lake City, and Santa Fe_

4. Which of the communities on the historical map are not on the contemporary map? What does this tell you? _Fort Hall and Bent's Fort have disappeared entirely_ from the map. Council Bluffs and Independence have become suburbs of larger urban areas.

83

Lesson 14

ACTIVITY   Compare and contrast two urban centers of the West.

Two "Hot" Spots

The BIG Geographic Question   What physical and human features have attracted people to two very different desert communities?

In the article you read about the extremes of the Mountain and Intermontane West. The map skills lesson helped you compare past and present travel routes to the West. Now learn what has made two specific communities of the Intermontane West important urban centers.

A. In the Almanac read about the settling of Salt Lake City, Utah, and Las Vegas, Nevada. Use the chart below to help you organize your answers to the following questions.

Question	Salt Lake City, Utah	Las Vegas, Nevada
1. When was it settled?	1847	1905
2. Who settled it?	Mormons	miners
3. Why did they settle there?	The Mormons were escaping religious prejudice.	It was established as a railroad town for the nearby mines.

B. Use what you know about the physical features and climate of the region to help you answer these questions.

Question	Salt Lake City, Utah	Las Vegas, Nevada
1. What problems might settlers of this location face?	• desert climate • scarce water • second harvest devastated by grasshoppers	• desert climate • scarce water
2. How did the people of the region overcome the problems?	• seagulls from lake ate grasshoppers • irrigation systems were set up using a reservoir and dams	• major irrigation projects created • irrigation projects led to the agricultural region in the west-central part of Nevada
3. Were these solutions successful?	yes	yes

C. Read about the communities today. Think about how they have changed since they were first settled.

Question	Salt Lake City, Utah	Las Vegas, Nevada
1. What things are important to the city today?	• state capital • center for the Mormon Church • mining • missile production industry • site of the next Winter Olympics	• entertainment • casinos
2. What role does tourism play in the community now?	• tourism will become more important with the Winter Olympics of 2002 planned there, but manufacturing is still its main industry	• tourism is extremely important; hotels, restaurants, and casinos are state's major source of income

D. Write a brief article comparing and contrasting the two urban centers of Salt Lake City and Las Vegas. To go with your article, draw a picture to show the differences in the styles of the two communities. (Make sure students' articles and images reflect the similarities and differences between the two cities.)

85

A. Study the physical map of California. Use the information on the map to complete the following.

1. Trace the line of the San Andreas Fault on the map.

2. How long is the San Andreas Fault? more than 750 miles long

3. What major cities are located on or near the fault? San Francisco, Oakland, Fremont, San Jose, San Bernardino, and Los Angeles

4. Write a brief desciption of where the fault is located, including where it begins and ends. The San Andreas Fault is located mainly on the western side of California. It reaches from near the Mexican border, up through San Francisco and Oakland, and into the Pacific Ocean.

B. Continue to study the map, particularly the San Andreas Fault. Complete the following.

1. Which California city has been hit by two severe earthquakes, one in 1906 and another in 1989? San Francisco

2. Use the chart below to describe the city, including where in California it is located and its land and water features.

City Name	San Francisco
Where It Is Located	toward northern California; between the Pacific Ocean and the San Francisco Bay; on the San Andreas Fault
Land and Water Features	hilly; almost completely surrounded by water

3. Do you think the city's location makes it more vulnerable to the activity of the San Andreas Fault than other cities located along the Fault? Why?
Yes, because it is the city most surrounded by water.

4. How does the San Andreas Fault affect life in California? It makes living in California dangerous because of the possibility of earthquakes and the destruction they cause.

89

Lesson 15

ACTIVITY Illustrate the history of water management in California.

Where's the Water?

The BIG Geographic Question What are some effects of water management on communities?

In the article you read about the diversity of California's human and physical features. The map skills lesson focused on the fragile San Andreas Fault in California. Now consider one of the state's other main concerns, the strained water supply. Create a model of California's physical relief to illustrate the location of the state's major efforts to reduce this strain.

A. California has four geographic regions. Find out about the regions. What are their water needs? Copy the following chart onto a sheet of paper. Complete the chart using the Almanac.

Region	Cities	Natural Features	Economic Activities	Source of Water
The Coast	• Los Angeles • San Diego • San Francisco	• Pacific Ocean • Coast Range	• entertainment • computers • fishing	• Sierra Nevada Range • Hoover Dam
The Central Valley	• Sacramento • Fresno • Stockton	• Sacramento River • San Joaquin River	• agriculture	• Sacramento River • San Joaquin River
The Mountains	• Lake Tahoe resort area	• Lake Tahoe • Yosemite National Park • Sierra Nevada • Coast Range	• recreation • lumber • mining	• rain • snowfall
The Deserts	• Palm Springs resort area	• Salton Sea • Mojave Desert • Death Valley	• some mining • recreation	• Colorado River • Owens Lake

---

B. Use the Almanac to find out about these projects and complete the chart.

Project	Starting Point	Ending Point
Los Angeles Aqueduct	Owens Lake	Los Angeles
Colorado River Aqueduct	Colorado River	Southern California cities and farms
Hetch Hetchy Aqueduct	Tuolumne River	San Francisco area

C. Make your own relief model of California. Mix two parts salt (2 cups) to one part flour (1 cup), and some tap water (3/4 cup) to produce a moldable clay. Mix in a bowl and add 2–3 drops of cooking oil. (Students' relief models should accurately indicate the location of urban cities, agricultural areas, and major water projects.)

1. Plan your map on paper, marking the locations of California's four geographic regions.

2. Transfer the clay mixture to the map. Mold the mountains and other geographic features in relief and allow your model to dry.

3. When the model is completely dry, paint the major regions different colors using tempera or poster paints.

4. Use the information you have gathered from the charts to locate and illustrate the following features on your model:
   • Major urban areas
   • Major agricultural regions
   • Major water projects (dams, aqueducts, and reservoirs)

D. Think about the role water has played in the growth of California's population and industries. What problems do you think California might have with water in the future? What solutions would you recommend?

(Students' answers should emphasize the limitations of natural resources, like water, and recommend conservation measures, such as not letting faucets drip, taking baths instead of showers, or only watering gardens during nonpeak hours.)

A. Study the map and the precipitation key.

1. What areas have the highest average precipitation? the areas along the Pacific Coast

2. What physical features are found in or near the areas that have the highest precipitation? mountains and rivers

3. What areas show the lowest average precipitation? the central and eastern parts of Oregon

4. What physical features are found in the areas that have the lowest precipitation? mountains, rivers, a desert, and a basin

5. Look back at the diagram of orographic precipitation in the article on page 92. Now look at the map of Oregon and the precipitation key. Draw the diagram onto the map in the appropriate place. (Make sure students' diagrams drawn onto the map include and label Pacific Ocean, Cascade Range, the desert, and plains.)

B. Look at the map diagram you have created and answer the following questions.

1. What happens to the air as it moves up the ocean side of the mountain?
The air gets cooler as it flows up the mountain.

2. Where does it rain? It rains on the ocean side of the mountain.

3. What is the air like as it moves down the far side of the mountain?
The air is drier. It gets warmer as it flows down the mountain.

4. How does this precipitation affect the timber industry in the area?
Areas that receive great amounts of rainfall provide adequate water that makes lots of trees grow.

Lesson 16

ACTIVITY Learn to see different sides of an environmental debate.

Depending on Forests

The BIG Geographic Question How can forest management issues be resolved?

From the article you learned about the vast amounts of timber in the Pacific Northwest and the benefits and problems connected with it. The map skills lesson showed you how the west side of Oregon gets more rain, creating forests on the western side of the state. Now investigate the debate over logging practices in the Pacific Northwest.

A. Answer the following questions about timber.

1. What are some items in your home or school that are made from timber?
(Possible answers include: pencil, paper, ruler, desk, house, box)

2. Where do you think the timber to make these items comes from?
forests or wooded areas

B. Imagine that the items you listed above depended on timber from the Pacific Northwest.

1. How does rainfall affect the growth of trees in this region?
The Pacific Northwest region gets heavy rainfall, which causes more trees to grow.

2. What is the most common method of harvesting the trees in the Pacific Northwest? clear-cutting

---

C. Many people disagree over whether or not to harvest trees in the Pacific Northwest. What are some of the advantages of cutting down trees there? What are some of the disadvantages? Put your answers on the chart below.

Advantages	Disadvantages
supplies lumber for products	creates erosion
supplies jobs for the area	destroys wildlife habitat
forest is replaced and can be used again in 70 years	destroys old-growth trees

D. Different groups of people feel strongly about the issue of cutting or not cutting down timber in the Pacific Northwest. Choose one of the following groups of people and think about how they would answer the questions below.

furniture makers

loggers

wildlife scientists

1. How does cutting down timber in the Pacific Northwest affect you?
(Students' answers should represent the perspective of the groups they chose and how cutting down trees affects them.)

2. What solutions would you suggest for the issue of cutting or not cutting down the timber?
(Students' answers should be logical and take on the perspective of the groups they chose.)

E. Write a proposal to the United States Secretary of the Interior about this issue. State your view based upon the group that you chose. Take into account how others feel on the issue. What kinds of solutions are there that everyone could agree to?
(Students' letters should reflect their chosen groups' viewpoints and take into account other viewpoints. Their solutions should be logical.)

A. Look at the map and answer the questions below.

1. List the nine islands that make up the state of Hawaii.

   a. Hawaii   f. Maui

   b. Kahoolawe   g. Molokai

   c. Kauai   h. Nihau

   d. Kaula   i. Oahu

   e. Lanai

2. Which color on the map represents the highest density, or concentration of people?

   purple

3. Which color represents the sparsely populated areas—areas with the fewest people?

   yellow

B. Urban means "in, of, or like a city." Rural means "in, of, or like the country." Use the map to answer the following questions.

1. Which map color would you say best represents urban areas? purple

2. Which color would you say best represents rural areas? yellow

3. How many people per square mile live on the northern tip of the island of Hawaii? 10 to 50 people per square mile

4. How do the coasts of the islands compare to the inland areas in population?
   In general, more people live on the coasts.

5. Which Hawaiian island has the highest population? Oahu

6. On which island are most of the cities located? Oahu

7. Compare the city locations with the population density and note your observations. Most cities are located along the coasts of the islands. These cities have the highest population densities.

101

Lesson 17

ACTIVITY   Discover how people have changed Hawaii's ecosystem.

People Causing Change

The **BIG** Geographic Question   How did people change the plant and animal life on Hawaii?

From the article you learned that people from many cultures settled on the Hawaiian Islands. The map skills lesson showed the population densities of the islands. Now find out how people have affected the landscape of Hawaii.

A. The ecosystem of Hawaii includes many living things that are not found anywhere else on Earth. Use the Almanac to find information about the unique life-forms of Hawaii, past and present, and about the effects people have had on the ecosystem.

1. List some of Hawaii's unique plants and animals below. Circle any that are endangered. (Students' answers should include the birds, plants, and animals listed on page 127 of the Almanac.

2. Why do you think the ecosystem of Hawaii is unique? (Students' answers should reflect an understanding of the effects of Hawaii's climate and its isolation from other land areas.)

B. Use what you have learned to complete the chart below. Tell how the climate affects each feature and how that feature affects the ecosystem. Add other features you have learned about.

(Possible answers include the following.)

Economic Activity	How Affected by Climate	Effect on Hawaii's Ecosystem
sugarcane plantations	Sugarcane needs 80–120˝ of rainfall and 75–86° temperatures.	Large plantations cleared trees, destroying the habitat of native plants and animals on the islands.
pineapple plantations	Pineapple needs warm, moist climate with well-drained soil.	Streams were shifted to support crops, which prevented fish from swimming and reproducing in the ocean.

C. Use the Almanac to find out where pineapples and sugarcane are grown in Hawaii. Then trace or draw a map of Hawaii and place symbols on the map to show the locations of the pineapples and sugarcane. Remember to draw a map key for your symbols.
(Pineapple symbols should appear on central Molokai, Lanai, and Oahu. Sugarcane symbols should appear on Hawaii, Maui, and Oahu.)

D. Write a few sentences explaining the effects of human actions on Hawaii's ecosystem.
(Students' answers should indicate an understanding of how human actions might alter Hawaii's physical environment. For example, rerouting water from one place to another interrupts and prevents life from thriving in the place the water is taken from. Or, the homes of animals are destroyed when the land is used for building.)

A. Look at the map and identify the following features.
(Make sure students have correctly identified the features on the map.)

1. Find an international boundary line and trace it with a red line.

2. Find the highway line and trace it with a yellow line.

3. Find the railroad line and trace it with a blue line.

4. Find the Trans-Alaska Pipeline and trace it with a green line.

B. Look at the legend on the map. Find these symbols and write what each one stands for.

1. ······· railroad

2. —— pipeline

3. -·-·- Dalton Highway

4. — river

C. Use the map to answer these questions about international boundaries.

1. What country borders the east side of Alaska? Canada

2. How can you tell the boundary line between Alaska and that country?
   A line separating the two places is drawn on the map, and they are different colors.

3. What country borders the west side of Alaska? Russia

4. What body of water forms a boundary between Alaska and that country?
   the Bering Strait

D. Think about the boundaries between Alaska and Canada and Alaska and Russia. Write a few sentences explaining which is a physical boundary and which is a human boundary. Explain your answer.
The boundary between Alaska and Canada is a human-made boundary because it is a line that only shows up on a map drawn by people. The boundary between Alaska and Russia is a physical boundary because it is a body of water formed by nature.

Lesson 18

ACTIVITY   Find out how communication and transportation can affect people.

Saving the Land

The **BIG** Geographic Question   How have the lives of native Alaskans been affected by transportation and communication developments?

From the article you learned that as more ways of traveling to and in Alaska came about, more people went there. The map skills lesson showed how different lines on a map represent features of Alaska. Now find out what happened with the land in Alaska.

A. Use the information in the article, map skills lesson, and Almanac to complete the following about Alaska.

1. Describe the land of Alaska?
   Alaska is very mountainous and remote. It has many rivers and lakes which remain frozen most of the year. Most of it is unsettled.

2. Who are the native people of Alaska?
   The Aleut and Inuit are the main native peoples of Alaska.

3. How do the native people usually travel in Alaska?
   The native people often travel by dogsled.

4. How has the land of Alaska been used to connect it to other places?
   People have built roads, a railroad, and a pipeline to connect Alaska to other places.

B. Review the information you learned about why the native Alaskans wanted to reclaim land and how the United States government responded to their claims. Organize the information you find on the chart below.

People	Actions Regarding Land in Alaska
Native Alaskans' claims	They considered Alaska to be their home. They believed they were losing control of their homeland. They wanted to govern themselves.
United States government's response	They used much of the land to build a pipeline to ship oil. They wanted to keep control of the oil rights in the north. They settled a claim with the Native Alaskans, giving them millions of acres of land and almost $1 billion.

C. Use what you have learned from the activity and the Almanac to role-play with a partner how you might have decided to divide the land and cash. Then write a short paragraph below describing your decision. Include a description of how your decision compares to what the Native Alaskans decided to do.
(Make sure students have indicated their decisions for dividing the land and cash and how their decisions compare to what the Native Alaskans did.)